# Introduction to Civil Engineering

A Student's Guide to Academic and Professional Success

*Revised First Edition*

By S.T. Mau and Sami Maalouf

*California State University—Northridge*

cognella
academic publishing

Bassim Hamadeh, CEO and Publisher
Michael Simpson, Vice President of Acquisitions
Jamie Giganti, Managing Editor
Jess Busch, Senior Graphic Designer
Marissa Applegate, Acquisitions Editor
Luiz Ferreira, Licensing Specialist



First published in the United States of America in 2015 by Cognella, Inc.





ISBN: 978-1-5165-5312-9

cognella
academic publishing

www.cognella.com 800-200-3908

# Contents

## CHAPTER 5: LEGENDS, MILESTONES AND LANDMARKS



## CHAPTER 6: ENGINEERING ETHICS



## CHAPTER 7: LIFE BEYOND THE BSCE DEGREE

# Preface

This book is written to inspire and empower students pursuing a Bachelor of Science in Civil Engineering (BSCE) degree. It is designed as a textbook for an introductory course in a civil engineering curriculum.

Students who come to a BSCE degree program may have some vague idea of civil engineering as a discipline or a profession. Some may have mistaken civil engineering as being mainly about architectural design. While it usually takes four years of study to learn to some degree what civil engineering is about, it is possible to provide a general description of civil engineering as a discipline by going through its major technical areas and the attributes of civil engineers. This is accomplished in Chapter 1.

Furthermore, civil engineering freshmen usually have no clue about what it takes to succeed in the freshman year and subsequent years. Chapter 2 gives a description on all the beneficial skills and tools needed to succeed as a civil engineering student.

Chapter 3 describes the common BSCE curriculum design. It also includes a description of the Fundamentals of Engineering (FE) exam, which contains many of the topics taught in the first three years of a BSCE curriculum and is usually taken by civil engineering students in their senior year. Chapter 3 also contains a description of the accreditation agency, ABET Inc., and its requirements for all engineering degree programs.

The importance of extracurricular activities to students' academic performance and progress is now recognized by many educators and deemed an integral part of student life. In Chapter 4, the most relevant student organizations on many campuses are described with emphases on student competitions and scholarship opportunities. The competitions sponsored by ASCE (American Society of Civil Engineers), the annual steel bridge competition and the annual concrete canoe competition, are described in some detail to encourage student participation.

With its long history, civil engineering is associated with many famous landmarks and monuments, and stories of those who built them. Modern civil engineering development is marked by key milestones and the efforts of pioneers. Chapter 5 gives an account of the landmarks and milestones of civil engineering and the human story behind them.

Engineering Ethics is a required subject according to ABET, yet it is often difficult to include in other civil engineering courses. A treatment of the subject and a discussion on student ethics regarding academic honesty and student behavior in and out of classrooms are given in Chapter 6.

After the BSCE degree, there are many career paths open to the degree holder. A typical path is to work as a civil engineer. Some high-profile private companies and public agencies who hire civil engineers are described in Chapter 7. Opportunities for advanced studies in civil engineering are also described, as well as opportunities outside of the civil engineering profession. The PE (Principle and Practice of Engineering) exam, the final step before obtaining a professional engineer license, is described in some detail.

# Chapter 1

# What Is Civil Engineering?

## 1.1 Overview

Civil engineering is the oldest engineering discipline. From the pyramids in Egypt, the Roman aqueduct and roads, to the great walls and the grand canal of China, ancient civil engineers left their imprint on human history on a grand scale. While the basic needs for civil engineering have not changed throughout the ages, the content and level of expectation of civil engineering work have certainly changed with the time because the tools available to civil engineers are changing with the time.

Civil engineering is a branch of engineering that deals with providing people with a livable built environment consistent with the standards and expectations of modern living through the applications of mathematics, science, and human experience. Some of the contributions of civil engineering are visible and obvious: buildings, bridges, highways, railways, airports, and dams and levees. Some are less commonly known as the product of civil engineering: offshore platforms, cell phone towers, power transmission lines and substations, drinking water and wastewater treatment plants, traffic signals, air pollution control, International Space Station, and many more. In short, civil engineering deals with people's everyday needs and more. The following table illustrates the connection between five basic human needs and technical areas within civil engineering.

Civil engineering's progress throughout time introduces new contributions to people's daily lives. Take the 20th century, for example; the U.S. National Academy of Engineering, a non-government, non-profit organization, after an elaborate nomination-and-review process published the twenty greatest engineering achievements of the 20th century in 2000. These are:

Among the twenty, clearly Number 4 and Number 11 are the contributions of civil engineering. Even Number 1, Electrification, the generation and transportation of electrical power, cannot be achieved without the civil engineering contribution to the designing and constructing power transmission towers and lines. Same is true for Number 2 and Number 3.

As already indicated in Table 1.1, civil engineering includes several very different technical specialty areas. By describing these technical areas, it is hoped that a clearer picture of what civil engineering entails may emerge. Eight civil engineering technical areas are described below. Each technical area has its own

Table 1.1 Basic needs and civil engineering

| Human Needs | Specific Nature of Needs | Civil Engineering Technical Areas |
|---|---|---|
| Breath | Clean air | Environmental Engineering |
| Drink | Safe water | Environmental Engineering |
| Sleep | Livable shelter | Structural/Construction Engineering |
| Move around | Ways to travel | Transportation/Construction Engineering |
| Safe from disaster | Earthquake mitigation | Structural/Geotechnical Engineering |
| | Flood mitigation | Hydraulic Engineering/Water Resources |
| | Wind mitigation | Structural Engineering |
| | Fire resistant | Structural Engineering |

sub-areas of specialty. They are described following the short overview of each technical area. When you begin to take civil engineering courses, you may identify each course with some of these technical areas. A student is not expected to be exposed to all the technical areas but at least four are included in the curriculum of any civil engineering degree program (see Chapter 3). Three contemporary issues confronting civil engineers are described following a description of other areas closely related to civil engineering. Personal

1. Electrification
2. Automobile
3. Airplane
4. Water Supply and Distribution
5. Electronics
6. Radio and Television
7. Agricultural Mechanization
8. Computers
9. Telephone
10. Air Conditioning and Refrigeration
11. Highways
12. Spacecraft
13. Internet
14. Imaging
15. Household Appliances
16. Health Technologies
17. Petroleum and Petrochemical Technologies
18. Laser and Fiber Optics
19. Nuclear Technologies
20. High-performance Materials

attributes of a typical civil engineer is portrayed near the end. The chapter ends with a brief mentioning of all the steps leading to a civil engineering degree and career.

## 1.2 Structural Engineering

Structural engineering is the technical specialty that deals with the analysis and design of constructed structures. From spacecraft to deep sea submarines, from tiny micro-electro-mechanical system (MEMS) devices to long bridges and tall buildings, these are all human-made structures that serve specific functions. A structure is always subjected to the many "loads" the environment forces upon it. These loads include the omnipresent gravitational load of its own weight (called the dead load), the weight of things moving about in or upon the structure (the live load), and event-driven loads originated from the occurrence of earthquakes, strong wind, or heavy snow. Structural design aims at providing a structure with sufficient level of resistance against these loads with minimum cost. Within structural engineering, there are several technical sub-areas. Some are named according to the type of structure. Some are named according to the type of load.

**Earthquake Engineering**. The suddenness of earthquakes and the damage they could cause in a matter of seconds inspired the study of the nature of earthquakes and the effects they inflict on structures. The effects of earthquake ground motion create vertical and horizontal forces that change violently within a short duration. The time-varying nature and the multi-directional nature of the earthquake-induced load require special design and analysis considerations. The fundamental approach in earthquake engineering is not to design a structure to withstand any earthquake at all costs but to design a structure that will not inflict injury to human lives at a reasonable cost.

**Wind Engineering**. Strong wind caused by a hurricane, a tornado, or a storm creates effects on structures that are also time-varying and multi-directional. Strong wind around a structure may push against a surface while creating a partial vacuum behind another structural surface. Unlike earthquakes, which occur infrequently, especially the damaging ones, strong wind in some areas occurs frequently and so is the damage it incurs. Design against such wind-related effects is the realm of wind engineering. Here again, the design approach is to protect human lives with a reasonable cost.

**Structural Reliability**. The many loads a structure must withstand during its life span are mostly of a "random" nature, meaning it cannot be defined precisely with respect to its magnitude and time of occurrence. So are the resistance provided by the size and material of structural components. Design in the face of uncertainty requires the application of probability and statistics. Structural reliability is the methodology applying these mathematical tools to the load-resistance analysis in structural design. It is used in the development of design codes and specifications that are followed by designers to provide acceptable levels of safety against all loads.

**Fire Engineering**. In the event of a fire in a building, the high temperature created by the fire may cause the building material to lose its strength and eventually fail under the weight of the building. Fire engineering in the context of structural engineering deals with the effective application of protective materials to the structural components such as steel beams and columns such that sufficient time is provided for the occupants to escape and the firefighters to arrive. The research in fire engineering provides data to be incorporated into design and construction codes and specifications.

**Bridge Engineering**. Some structural engineers specialize in bridge design and construction. Bridge design can be categorized according to material and bridge type. One unique feature of bridge design is it is closely integrated with construction. From the bridge foundation to the superstructure, the process of construction and erection often requires detailed analysis by the design engineers and likely dictates the designers' choice of bridge type.

**Dam Engineering**. The design of dams requires detailed study of the geological characteristics of the site and the mechanical properties of the foundation before the dam type is selected. For some types of dams, it is necessary to ensure the dam material is placed in such a manner that seepage of water through and under the dam body is within acceptable limits. Dam engineers also design all details on how to divert water during construction and specify maintenance and operations procedures post construction.

**Building Engineering**. Structural engineers often become building design specialists because building design is more frequently in demand than bridge or dam designs, especially in urban centers. Building engineers also design special buildings such as stadiums and large dome structures.

**Forensic Engineering**. Forensic engineering refers to the study of causes of an engineering event, usually a disaster or failure of some kind. In the context of structural engineering, it refers to the investigation of a structural failure. There are no courses or programs for structural forensic engineering training, but experienced structural engineers who have investigated past failures are often called upon to investigate a new event. In case of major disasters, often a team of experts are assembled to study the cause of the disaster and to make recommendations to prevent future disasters. Even when the cause of disaster is terrorism, forensic engineering would reveal the weakness in design and provide guidance for future designs. A good example is the 1995 bombing of the Alfred P. Murrah Federal Building in Oklahoma City. The collapse of the building was caused by the bombing that destroyed ground-level columns in the front of the building. Experts recommended that future buildings should have sufficient redundancy in the design of supporting columns so that the damage of one or more columns would not lead to the collapse of the whole building.

In structural design in the context of civil engineering, there are three construction materials that are dominant: steel, reinforced concrete, and timber. Each has its own design specifications. Thus, steel structure design, reinforced-concrete structure design, and timber structure design are three main design disciplines.

## 1.3 Geotechnical Engineering

Most civil engineering structures are earthbound. They sit on soil and rock ground directly or on constructed foundations that transfer the load to the soil or rock below. Geotechnical engineering is the technical specialty that deals with soil and rock as supporting materials for structures. It deals with the various foundation types that work between the structure and the ground. In addition it deals with the stability of soil or rock slopes whose failure may cause loss of human lives or damage to property. There are several technical areas of study that are pertinent to geotechnical engineering.

**Engineering Geology**. While geology is a basic science that is concerned with macroscopic earth structures or movements, engineering geology provides geological data pertinent to constructed structures. One obvious example of the application of engineering geology is the mapping of active seismic faults that are to be avoided when making plans for human habitat development, roadway construction, or power plant construction. At a more fundamental level, understanding various geological formation and rock types provides geotechnical engineers the knowledge necessary in assessing the suitability of a site for human activities.

**Soil Mechanics**. Most people would not consider soil as an engineering material, but it is, because most constructed structures are situated on it by necessity. Without due consideration of soil's bearing capacity under various circumstances, a structure built over it may sink, tilt, or outright turn over. Soil mechanics is a branch of mechanics that studies the mechanical properties of various types of soil and its strength at different moisture-content levels. It provides the scientific base upon which design formulas and codes are developed for everyday engineering design practice.

**Rock Mechanics**. The properties of rock become relevant when it is used as the foundation of a high-rise building or a large dam. It is also relevant when one examines the stability of the slope of a mountain or a tunnel. It is also the subject of study for the occurrence of earthquakes.

**Foundation Engineering**. A foundation is the interface between a superstructure and its supporting soil. A common type of foundation for single-family homes consists of strip footings placed under load-bearing basement walls. Another common practice is to use a concrete slab to spread the weight of the building over the soil underneath. Foundation engineering is the study of different types of foundation and their proper applications. Depending on the properties of soil at a site, shallow or deep piles may be deployed. The construction of a bridge over water may require the use of deep caissons on which piers are constructed. To stabilize an excavated slope, various types of methods may be used including retaining walls and slope-protection vegetation growth.

**Soil Improvement**. When a structure must be placed at a site with very weak soil, various techniques can be used to improve the soil properties. These typically involve the use of replacement material through excavation or the injection of special material (grouts) into the original soil to change its properties. Another special technique is to place geo-synthetic fabrics or textiles in horizontal layers to strengthen the soil or to limit soil's permeability, which is essential in the design for landfill and hazardous material deposit sites.

**Tunnel Engineering**. Tunneling through soil or rock is sometimes necessary in the construction of roadways or special storage spaces. Tunnel engineering deals with the route determination, selection of tunneling machines, and the analysis and design of the tunnel structure.

Most of the things designed by geotechnical engineers are not as visible as those by structural engineers because they are underground or under the superstructure above. But, it is safe to state that no civil engineering work can be constructed without the contribution of geotechnical engineers.

## 1.4 Environmental Engineering

Environmental engineering is the application of engineering means to protect human health and to preserve the natural environment by managing and developing water, air, and land resources. The application of environmental engineering relies on the fundamental sciences of chemistry, biology, ecology, and health sciences. Most modern environmental engineering projects are planned and implemented under the auspices of the Clean Water Act, Safe Drinking Water Act, Clean Air Act, and other federal and state environmental legislation. Several technical areas in environmental engineering are described below.

**Water Treatment and Supply**. Before water is consumed, it has to be collected first from either underground or above-ground sources. Therefore, source control is one of the most important tasks of water supply. Except for a few municipalities where the source water derived from deep aquifers, source water has to be treated to remove contaminants such as pathogenic bacteria, heavy metals, and pesticide residues. The process of treatment involves the removal of suspended solids and the use of chemicals or ultraviolet (UV) radiation to disinfect unwanted organisms so that the effluent water satisfies quality requirements dictated by the Federal Safe Drinking Water Act. For water used by industrial plants such as paper mills or nuclear power plants, special treatment is needed and its discharge is regulated.

**Wastewater Treatment and Disposal**. In a modern municipality, household wastewater is collected through underground pipelines to a treatment plant. The wastewater treatment process is very different from drinking water treatment and is classified into primary treatment, secondary treatment, and tertiary advanced treatment. Primary treatment removes suspended solids from wastewater by a sedimentation process. Secondary treatment is to remove dissolved organic wastes from wastewater by biochemical decomposition followed by further sedimentation. The Federal Clean Water Act establishes nationwide minimum treatment requirements for all wastewater. For municipal wastewater discharge, the minimum treatment is the secondary treatment, which removes 85% of biochemical oxygen demand (BOD) and total suspended solids (TSS). BOD is a measurement of oxygen-demanding organic wastes. In situations in which these minimum treatment levels are not sufficient, the Clean Water Act requires additional treatment, which is accomplished by membrane filtration and other physical-chemical processes.

The outcomes of the wastewater treatment are solid sludge and effluent water. The solid sludge sometimes can be used for landfill or even as fertilizers. The effluent water can be used for irrigation or groundwater recharge or may be directly discharged into river, stream, or lake or sea. For a municipality, the amount of rainfall determines the ways of collecting and treating wastewater. Obviously if a large amount of rainfall is expected, especially when storm water comes in a very short period of time, rainwater runoff should be separated or diverted either temporarily or permanently from household wastewater in order to avoid overwhelming the treatment plant. Thus, a wastewater collecting system can either be combined (for more arid areas) or separated. Some industrial plants produce special wastewater that requires the removal of heavy metal or hazardous chemicals before being discharged.

**Air Pollution**. Environmental engineers monitor, analyze, and assess the air quality around municipalities. Air pollution comes from natural and human-activity sources. Volcanic eruption is a major natural source of air pollution. The gaseous and particulate contents of a volcanic eruption are often studied by scientists rather than engineers. Around a large municipality, however, air pollutants come from automobile emissions, nearby industrial plant emission, and even from faraway sources. Health science advances have discovered that tiny solid particles in the air such as soot are hazardous to human health. The monitoring of these particles is as important as that for gases. Tracing plant emission in the atmosphere, called plume analysis, is important in the assessment of the environmental impact of a plant. Another form of air pollution is sand storm. Monitoring of sand storms may lead to the sources of the storm and policy for conservation or planting of new vegetation.

**Solid Waste Disposal**. Solid wastes, commonly known as trash and garbage, from domestic, commercial, and industrial sources are to be collected, separated, and partially recycled, and disposed of in landfills and special disposal sites. Environmental engineers, working with other civil engineers, select, design, and construct sanitary landfill sites. Water percolating through a sanitary landfill is intercepted, collected, and treated in order to prevent the seepage of hazardous materials into ground water strata. Some solid waste maybe burned by specially designed incinerating plants.

**Nuclear Waste Disposal**. Nuclear waste comes from used fuel rods in nuclear power plants. Though the degree of radiation from these spent fuel rods is low, long-term exposure to low-level radiation is hazardous to human health. Disposal of these wastes has few options. The basic approach is to store them in places far from human habitat. Furthermore, it must be assured that the storage containers will not leak to the environment in any way. Leakage to underground water would be disastrous because the contaminants can travel far and reach sources for human water consumption.

**Noise Pollution**. In modern municipalities, human activities often generate sustained high levels of sound that are hazardous to the physical and mental wellbeing of habitants. Sound barriers are often needed to shield neighborhoods from highway traffic noises. Power plants or air-conditioning plants on large campuses produce high levels of noises that also require containment and shielding. Environmental engineers monitor the noise levels and design and implement mitigation strategies.

**Environmental Impact Assessment.** Environmental engineers are often called upon to assess the impacts on human health and the natural environment by a new development, a new industrial plant, or even a new commercial establishment such as a large shopping mall. Such assessment may entail the study of noise, traffic, water consumption and discharge, power requirements, air pollution potential, and other factors.

Environmental engineering as a part of civil engineering is unique in its extensive applications of knowledge from health sciences and biology and chemistry. Its practice is also very much impacted by environmental laws enacted at the state and national levels.

## 1.5 Water Resources Engineering

Water resources engineering is a specialty dealing with the use of water in support of modern living, including the agricultural, industrial, domestic, recreational, and environmental needs. Its scope includes the finding and preservation of above- and underground water sources, understanding the movement of water in nature, engineering the transport of water, and managing erosive effects of water wave and current on shorelines. Some core and related specialties are described below.

**Water Resources System Engineering.** The understanding of the circulation of water on earth and managing the sources of water in a region requires a system approach. Decision on the water supply for a city or a region requires the knowledge of water sources and the quality and quantity of each source. The application of system analysis in water resources management and the design and operations of multipurpose reservoir and river systems is at the core of water resources system engineering.

**Hydraulic Engineering.** Hydraulic engineers design artificial waterways such as canals, channels, and aqueducts as well as manage water movement by designing and constructing dams, levees, canal locks, and other water-regulating devices. For many regions a major task of hydraulic engineers is flood prevention and control, which entails the assessment of potential rainfall quantity, prediction of water levels along natural rivers, streams, or channels, and strategies to mitigate flooding hazards by improving the natural topography. Hydraulic engineering is also fundamental to hydraulic-power generation. In hydraulic-power generation a prerequisite is a high water-level differential (water head). When water moves from a high level to a lower level, the difference in the water levels provides the energy potential for power generation. Some dams are constructed mainly for power generation although usually a dam also has the potential to be used for flood control. The stored water in a dam's reservoir can be used for agricultural, industrial, and domestic consumption as well as recreational sports.

**Coastal Engineering.** The movement of water in oceans and lakes has erosive effects on their shorelines. The preservation of wetland for flood mitigation or marine ecology requires the knowledge of such effects. Use of artificial barriers such as breakwaters or dikes at a shore or a harbor can result in reducing the water wave level within protected areas, eliminate or reduce the effects of shoreline erosion, and redirect natural sediment so that new land can be created over time.

**Ocean Engineering.** Ocean engineering deals with the effects of ocean currents and waves on ocean-bound structures and the analysis and design of such structures to withstand the wave forces. The most prominent ocean-bound structures are offshore platforms for oil exploration and production. Ocean engineers provide estimates of forces generated from waves and currents and the interaction of wave and structure so that structural engineers can design a platform to withstand such forces. Other ocean-bound structures include offshore wind farms and pipelines to transport materials from offshore to shore. While ships are obvious ocean-bound structures, their design usually falls in the realm of naval architecture, which integrates several engineering disciplines: structural, ocean, mechanical, and electrical for the design of ships. Naval architecture is not considered as a part of civil engineering.

## 1.6 Transportation Engineering

Transportation engineering deals with the efficient transport of people and goods. The content of transportation engineering changes whenever a new mode of transportation becomes viable. For example the advent of airplanes and air travel led to new technical fields such as airport design and air traffic control. Several sometimes overlapping technical specialties are part of transportation engineering:

**Transportation Planning**. Transportation infrastructure is mostly government funded or at least government approved. Before any physical facilities are designed and built for moving people and goods, decisions must be made from policy and political considerations. Transportation planning considers policy formation processes, cost, financing, and projected performance of potential transportation systems, including inter-modal transportation that involves more than one mode of travel such as sea-land-air travels.

**Transportation System Engineering**. Transportation System Engineering entails the efficient management and operation practices, design, and assessment of the cost-effectiveness of transportation systems. The assessment of transportation systems requires performance modeling techniques, traffic simulation, and environmental impact (noise and air pollution) analyses.

**Highway Engineering**. Highway engineering focuses on the planning, design, construction, and operation and maintenance of highways. Unique to highway engineering is the design and construction of highway pavements and foundations, and the design of highway interchanges. The operation of highways includes the use of high-occupancy lanes and networked signals and displays that can alter lane direction during rush traffic and warn travelers of road conditions ahead. Design and construction of toll booths and ways of collecting tolls are part of highway engineering as well.

**Railway Engineering**. With the advent of high speed railway, light-rail systems, and magnetic levitation systems, railway engineering gained renewed interest in civil engineering. Railway transportation remains a cost-effective way of transporting large quantities of goods on land. Railway engineering focuses on the planning, design, construction, operation, and maintenance of railways. Advances in electronic signal design and communication technology provide new tools in the control of railway traffic for efficiency and safety.

**Port and Harbor Engineering**. Even under the most favorable natural conditions, a port requires additional engineering to make sure ships can safely navigate through and dock, and cargoes can be efficiently unloaded and uploaded for shipping elsewhere. Some harbors require routine dredging to maintain the navigation channel. Some harbors need breakwaters to tame the ocean waves. These are the scope of port and harbor engineering.

**Airport Engineering**. The construction of a new airport requires extensive planning, including the study on demand, environmental impact, cost analysis, and investment returns. Since an airport is part of an air transport system, the impact of a new airport must consider the regional air traffic demand. Airport site selection requires extensive study of regional topography, including local constructed structures, prevailing winds, and, for airplane safety, movements of birds. The impact on nearby neighborhoods includes the new land traffic generated by travelers and the inevitable noise generated along air traffic routes. The engineering of airport infrastructure such as runways, terminals, and signals falls in the domain of other civil engineering specialties.

**Traffic Engineering.** Traffic engineering sometimes is considered as synonymous to transportation engineering but it is usually defined as the narrower field of management of traffic flow. From huge metropolises to small towns, surface traffic must be controlled and modulated for safety and speed. Traffic engineers use projected and monitored traffic patterns and volume to design automated or centrally controlled street signals to modulate traffic. Tools used for traffic control include weight sensors for triggering of left-turn signals and ramp-entry signals for freeway entry during rush hours.

**Intelligent Transportation Systems.** New and emerging electronic and computer technologies make it possible to efficiently monitor, evaluate, and improve the performance of transportation systems. An example of the applications to new vehicle engineering is the possibility of high-speed automated highway travel by automobiles in groups. Another example is the centrally controlled traffic monitoring and management systems used in large cities. Fully automated automobiles without drivers are in development and testing.

Transportation engineering is a specialty whose content is changing with emerging technologies. The operational aspect of transportation engineering is heavily influenced by the advances in new technologies.

## 1.7 Construction Engineering

Transforming design details on paper or a computer file into physical reality is the task for construction engineers and managers. The successful completion of a construction project requires the integration of several areas: human resources management; financial resources and cost management; construction processes; schedule design and control; construction machinery; electric and mechanical facilities; legal, health, and safety issues; and risk management. Construction firms may specialize in one or more types of structures: buildings, bridges, dams, highways, airports, and ports and harbors. In the following some of the engineering and management aspects of construction engineering are described.

**Construction Processes.** Depending on the types of structures to be constructed and the materials to be used, each project has a unique process to follow to erect the structure step by step. The execution of a physical construction process often requires the knowledge of geotechnical engineering, structural engineering, construction materials, and site surveying. For example the construction of a multi-story building starts with the placement of foundations, which may consist of multiple steel or concrete piles. The story-by-story erection process depends on whether the building is of concrete or steel construction. These processes follow well-developed practices deeply rooted in accumulated experience of construction engineers. Intertwined with the construction process is construction scheduling, which lays out in detail the daily activities to be performed. Creating a construction schedule must consider the critical phases that determine the time length of the project.

**Electric and Mechanical Facilities.** In most construction projects electric and mechanical facilities are to be installed. Basic knowledge of common electrical and mechanical facilities is required for the installation of these facilities and their integration into the main structure.

**Construction Machinery.** From conveyer belts and bulldozers to scrapers, excavators, loaders, graders, compactors, cranes, and pipe-layers, many different machines are used in construction. Knowledge of these machines' capabilities is required in planning the types and quantity needed for a construction project.

**Financial and Cost Management.** From the bidding for a project to the actual execution, financial management is at the center of concern of construction management. Bidding is the critical process by which potential contractors compete to earn the contract from the owner of the project. Cost estimating is done at the beginning of a potential project and updated throughout the project. During the project period, cash flow management is needed to keep the project going. Successful contractors often have sufficient cash reserve or credit to ride through tough periods of low income cash flow and high outward cash payments.

**Contracts and Specifications.** A contract is a legal document that specifies the responsibilities of both the contractor who delivers the service and the owner who receives the service. Specifications are also legal documents that the contractor must follow throughout the project. Specifications are often based on well-developed construction practices but sometimes may include special requirements added by the owner. These legal documents are developed, negotiated, and decided upon between the contractor and the owner.

**Health and Safety Issues.** The health and safety of not only construction workers but also neighboring residents or passersby is regulated by the government. The Occupational Safety and Health Administration (OSHA) of the Department of Labor enforces regulations and laws pertaining to construction. Knowledge of these regulations and the strict adherence to the regulations are the responsibilities of the construction engineers or managers.

**Legal Issues and Risk Management.** Examples of potential liabilities of a construction project include negative cash flow, missing construction time milestones, mistakes in the constructed configurations or the strength of materials used, and accidents resulting in human injury or death. The prevention of any of the above and the timely management of the damage once any of them has occurred are the realm of risk management. Legal and ethical guidelines are to be followed at all times.

## 1.8 Geomatics (Surveying Engineering)

Geomatics is the new name for an expanded technical area of what used to be called surveying or surveying engineering. Surveying engineering is the technical specialty that applies the science of measurement to the assembling of spatial data on land and sea and any natural or constructed objects thereof. Its applications often involve legally required documentation related to the transaction of property, the location of routes or points needed in construction, and the collecting of global data for resources analysis and utilization. Some of the technical areas in surveying engineering are described herein.

**Plane Surveying.** The earth is a near circular sphere with a curved surface. When the area of interest on the earth's surface is small enough such that the curvature throughout the area can be neglected in measuring and locating any point in the area without causing any significant amount of error, the earth surface is treated as if it is a plane. With this simplification, the measuring and locating of points or routes are mathematically much simpler. Plane surveying deals with the techniques and skills in carrying out such tasks within the limit of the simplification. In plane surveying, optical instruments are used in conjunction with basic mathematics to accurately locate points and lines, to calculate areas and volumes, and to map existing surface features. Since no scientific measurements can be made without error because of limited resolution of instruments or even human mistakes, techniques are developed to minimize measurement errors.

**Route Location**. For highway or railway construction, the roadway configuration contains straight lines and curves in the horizontal and vertical planes necessitated by the change of natural topography and the need to minimize the cost of changing the natural topography. These lines and curves are designed and documented on paper or in computer files. Route location is the technique to transform points on these lines and curves on file to points on location with reference to some existing points on the earth's surface.

**Land Surveying**. Land surveying pertains to the measuring and documentation of property lines, streets and alleys, subdivisions and lots within a city or a village. The map of the area is called a plat, which forms the legal document in ownership transactions.

**Geodetic Surveying**. In contrast to plane surveying, geodetic surveying includes the earth's surface curvature in its calculations and mapping. More advanced mathematical tools such as spherical geometry and curvilinear coordinate systems are utilized for the computation of measure data and locating of points and lines. Surveying instruments may include the use of aerial photography, satellite imaging, and the global positioning system (GPS). Geodetic survey is necessary in the construction of long-span bridges and other structures covering a large stretch of area (a long tunnel, for example).

**Aerial Photography and Satellite Imaging**. With the advances in optical and electronics technologies resulting in improved measurement resolution and accuracy, images produced from airplane- or satellite-based instruments can be used in the mapping of ground features for various applications. Advances in digital photography and computer automation make the mapping efficient and much less time consuming. Applications of such images and maps range from real estate information generation and agricultural resources evaluation to military and intelligence information evaluations.

**Geographical Information System**. Surveying engineering provides the basis for the geographical information system (GIS) of a state, city, town, or smaller unit. Useful information such as basic population data, infrastructure, average income of population, demographic distribution, educational and recreational resources, crime rate, etc., is compiled in a single database for easy retrieving and application. GIS is also an example of the increasing application of information technology to surveying engineering.

It should be mentioned that surveying nowadays is often considered a technical area outside of civil engineering or engineering in general. After all, the organization in charge of testing for professional licensing in engineering and surveying, National Council of Examiners for Engineering and Surveying (NCEES), has separate exams for engineers and surveyors. Some civil engineering programs do not require any surveying courses.

## 1.9 Urban Planning

Urban planning integrates land use planning, infrastructure planning, and public policy for new developments or renewal of urbanized communities. Successful urban planning requires the application of the knowledge developed in social, economic, architectural, and engineering studies. Educational programs for future urban planners can be found in Bachelor of Arts degree programs in arts, architecture, and public policy colleges. Bachelor of Science degree programs are also available in some universities. The application of civil engineering to urban planning emphasizes the physical aspects of urban planning: street patterns, park and recreational areas, industrial and residential areas, transportation systems such as freeway access, airport access, public works and utilities, infrastructure management, etc.

## 1.10 Related Disciplines

Some of the disciplines that interact or overlap with civil engineering are described here.

**Applied Mechanics.** Mechanics is one of the early and basic areas in physics. It studies the nature and effects of force. Applied or engineering mechanics emphasizes the application aspects of the theory of mechanics. Civil engineering structures are subjected to the effects of gravitational force, wind, and earthquakes, and effects of temperature change. Various applied mechanics areas ranging from the fundamental statics, dynamics, fluid mechanics, and mechanics of materials to more advanced areas such as thermal stress and wave propagation have direct applications in structural engineering, geotechnical engineering, water resources engineering, environmental engineering, and construction engineering.

**Architectural Engineering.** Architectural engineering specializes in the aesthetics and structural design of buildings. The structural design aspect of architectural engineering overlaps with structural engineering. The aesthetical design aspect of architectural engineering applies the knowledge developed in architecture studies. Architectural engineers also study and apply knowledge in electrical and mechanical systems to building designs.

**Agricultural Engineering.** Agricultural engineering traditionally entails two technical areas: irrigation engineering and mechanization. Irrigation engineering is part of hydraulic engineering while the mechanization of agricultural engineering is part of mechanical engineering. Since the advent of DNA engineering, the use of basic molecular biology techniques to change the properties of a specific crop becomes an important addition to agricultural engineering and a new and broader field of bioengineering emerges.

**Aerospace Engineering.** Aerospace Engineering entails aeronautical engineering and astronautical engineering, which develops vehicles that fly through the earth's atmosphere and beyond, respectively. The structural analysis and design of the flying vehicles, airplanes, and spacecraft, is most similar to that of civil structural engineering. The difference is in the nature of forces acting on the structures.

**Biomedical Engineering.** Biomedical engineering specializes in the applications of engineering to the medical field, including medical imaging, surgical devices, and implant devices. The structural analysis and design of medical devices and implants overlaps with civil structural engineering in the basic tools used and differs in the nature of forces acting on the devices.

**Naval Architecture.** As briefly described before, naval architecture specialized in the design of ships. The analysis and design of ship structures is similar to that of civil structures and uses similar computational tools. The difference is in the nature of forces acting on the structures.

## 1.11 Civil Engineering Processes

No matter what kind of civil engineering project is involved, all civil engineering projects go through four main phases: planning, design, construction, and maintenance/operation. These phases sometimes are intertwined and overlapped.

**Planning**. As any civil engineering projects, private or public, involve public interest and major funding, the planning phase could sometimes run into many years. The very first stage of the planning phase is a feasibility study, which usually includes not only a financial study but also a study of legal issues. Virtually all major civil engineering projects must go through environmental impact studies and public hearings. For these studies and hearings, at least a preliminary design must be prepared and presented.

**Design**. At least two stages are involved in the design phase: preliminary design and final design. Sometimes it is necessary to have an additional intermediate design stage. The preliminary design is to create an outline of the concept, scope, structure, materials to be used, method of construction, and cost and timeline estimate of the project. The preliminary design can be part of the planning phase. The final design includes all detailed designs of every structure involved in the project and every associated facility such as electric and mechanical facilities.

**Construction**. The actual construction phase includes the physical erection of all the structures and in the meantime the observation of all applicable safety and environmental regulations during the construction phase.

**Maintenance**. When the construction phase ends and the owner takes over the project, the maintenance/operation phase begins. The owner of the project usually takes over all responsibilities, but the contractor is usually bound by a warranty agreement. During the warranty period interaction between the owner and the contractor could be very frequent and intense. Beyond the warranty period, the physical structures require constant maintenance. A well-known example is the constant painting of the cables of the Golden Gate Bridge in San Francisco.

## 1.12 Contemporary Issues: Civil Infrastructure Renewal

Infrastructure is the underlining fabric that makes a society function properly. Civil infrastructure refers to the underlining structures or facilities that are the product of civil engineering. These civil infrastructures can be grouped into the following categories, according to ASCE (The American Society of Civil Engineers):

Transportation: Aviation, Roads, Bridges, Transit, Inland Waterways.
Environment: Drinking Water, Wastewater, Solid Waste, Hazardous Waste.
Energy: Power Plants, Electric Grids.
Education: School Buildings.
Public Safety and Recreation: Dams, Levees, Public Parks and Recreation

As the United States is a developed country, many of the above civil infrastructures have been in service for 50 years or more. Most physical structures require maintenance to avoid breaking down or outright failure. The American Society of Civil Engineers evaluates the civil infrastructure status every four years since 2001 and publishes an American Infrastructure Report Card. The 2013 Report Card, as in previous years, assigns a letter grade (from A to F) for each of the infrastructure types as shown below:

**ASCE 2013 Grades (with the 2009 Grades in Parentheses)**
America's Infrastructure cumulative GPA: D+ (D)

ASCE also estimates the investment need in civil infrastructure is $3 trillion by 2020.

Renewal of America's civil infrastructure obviously will cost a large sum of money. ASCE's 2005 estimate for the five-year investment need was $1.6 trillion. By 2009 it became $2.2 trillion and by 2013 it

| | |
|---|---|
| Aviation | D (D) |
| Bridges | C+ (C) |
| Dams | D (D) |
| Drinking Water | D (D-) |
| Energy | D+ (D+) |
| Hazardous Waste | D (D) |
| Inland Waterways | D- (D-) |
| Levees | D- (D-) |
| Ports (New Category) | C |
| Public Parks and Recreation | C- (C-) |
| Rail | C- (C+) |
| Roads | D- (D) |
| Schools | D (D) |
| Solid Waste | C+ (B-) |
| Transit | D (D) |
| Wastewater | D (D-) |

is $3 trillion. It becomes obvious: the longer we wait because of cost and funding concerns, the more will it cost later.

Civil infrastructure renewal also needs different technology than just new construction. Replacing or repair leaky wastewater pipes may be accomplished with micro-tunneling methods that require no lengthy open trenching. Finding leakage in drinking water and wastewater pipes may utilize new optical or electromagnetic wave sensors. Other monitoring technologies for civil infrastructures are being developed.

## 1.13 Contemporary Issues: Sustainable Engineering

Sustainability is a concept that gradually gains recognition and is being put into practice in many sectors of the global economy. The 1983 Brundtland Commission (World Commission on the Environment and Development, WCED) of the United Nations released its report *Our Common Future* in 1987, defining sustainable development as: "development that meets the needs of the present without compromising the ability of future generations to meet their own needs." To be practical sustainable development must meet the needs of **social, environmental, and economic** concerns. A **bearable development** is one satisfies environmental and social concerns. A **viable development** is one satisfies environmental and economic concerns. An **equitable development** is one satisfies social and economic concerns. A **sustainable development** is one satisfies all three concerns. This is called the "triple bottom line," and is best illustrated by the following diagram.

Not surprisingly civil engineering projects are at the center of the sustainable development concept and practice. The key concepts are endurance, energy efficiency, and reduction of waste. Of the major sustainable development practice categories, several are civil engineering related:

**Water Conservation and Water Quality**. Once water rationing, even just for lawn use, became reality in southern California in 2009, water conservation was no longer a theory embraced only by environmentalists. Many practices for individuals and homeowners, such as limiting showers to five minutes and using more efficient toilets, can reap significant savings in water consumption. For civil engineers, using moisture sensors to regulate watering lawns is an excellent technical solution to water conservation. Another major technical challenge in water conservation is the reduction of leakage from water-supplying pipes. These leaks occur from both water mains and from the link between water mains and homes. The latter is the responsibility of homeowners. Unless cost-effective new technology is developed, large-scale improvement is unlikely because of the financial burden on homeowners. Drinking water quality is often taken for granted in America, but a 2009 report claiming 40% of people were exposed to dirty drinking water raised alarms.

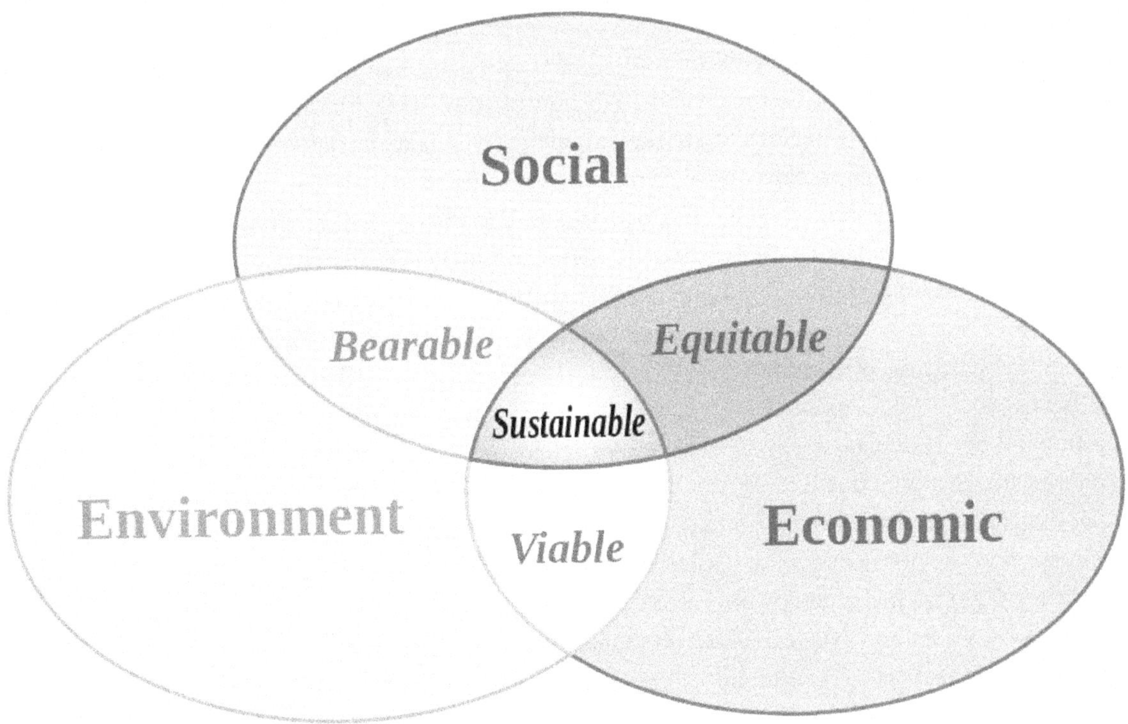


Figure 1.1 A sustainable development satisfies social, environmental, and economic concerns.

**Wastewater Collection and Treatment**. Wastewater collecting pipes are prone to chemical and biological corrosion especially in warm weather. Leakage from wastewater pipes creates environmental hazards that are difficult to mitigate. Thus cost-effective corrosion-resistant pipes that are more enduring are needed. Wastewater treatment uses large amount of chemicals and results in large amounts of sludge. Recycling sludge cake as fertilizers has been in practice for years. More energy-efficient and cost-effective treatment methods are much in need.

**Green-Building Technology**. Buildings, large or small, commercial or residential, are major sources of energy consumption. Building owners are increasingly aware of the ever-increasing cost for building maintenance and operations. They are taking steps to curb energy consumption and increase sustainability. The owner of the Empire State Building, opened in 1931, announced in April, 2009 that it was spending $20 million to enhance environmental efficiency by cutting energy use by 38 percent. The owner of one of the tallest buildings in the world, TAIPEI 101, opened in 2004, announced in November 2009 that $1.9 million was to be spent over a 20-month period in a transformational effort to reduce energy consumption by 10%, and water usage by 10%. The energy cost saved was estimated to be $600,000 a year. TAIPEI 101 planned to apply for the gold certificate from the **LEED** certification (see below).

The LEED certification is a consensus process that standardizes the recognition of energy-efficient buildings.LEED stands for Leadership in Energy and Environmental Design; it is pioneered by the **U.S. Green Building Council** (**USGBC**), which was founded in 1993 and is a non-profit trade organization promoting sustainability in how buildings are designed, built, and operated.

The LEED certificate system offers four levels of recognition: Certified, Silver, Gold, Platinum, with a 100-point scoring system. Factors considered for new constructions includesustainable sites (construction pollution, alternative transportation, site development, storm water, heat effect, light pollution), water efficiency (water-efficient landscaping, innovative wastewater technologies, water-use reduction), energy and atmosphere (optimize energy performance, on-site renewable energy, refrigerant management, green power), materials and resources (storage and collection of recyclables, building reuse, construction waste management, material reuse, regional materials, rapidly renewable materials, certified wood), indoor environmental quality (tobacco smoke control, ventilation, low-emitting materials, thermal comfort, daylight, views), innovative design, and regional priority. Factors for existing buildings are similar but take into account that only certain factors can be changed once a building is already built.

**Solid Waste Management**. Commonly known as trashes and garbage, industrial, commercial, and domestic solid wastes are becoming increasingly difficult to manage, not only because of ever-increasing volume but also because of ever-increasing variety and new waste types. The best strategy in waste management is to reduce waste. For example in late 2009, Japan promoted the use of reusable cups for drinks. Recycling is the next best approach. Civil engineers design and build solid waste distribution centers, where incoming garbage is sorted and redistributed. The last approach is to safely transport and deposit garbage in well-designed and -constructed landfills. As landfill materials often generate gas, which can be used to generate power, use of micro-turbines at landfill sites to generate power for onsite power needs is gaining momentum. Civil engineers design and construct safe and durable landfill sites.

**Hazardous Waste and Nuclear Waste Management**. Hazardous waste from industrial and domestic sources as common as battery cells and computer monitors cannot be simply dumped into landfill sites. Materials contained therein, most commonly heavy metals, are regulated materials and must be processed and recovered safely. Civil engineers design and construct treatment plants. For nuclear waste from nuclear power plants, defense establishment, and medical usage, the only solution now is to safely stow it away from population centers. Because of the long life of radiation materials, storage sites are to be selected for long-term geological stability and safe distance from underground water sources. The selection, design, and construction of such sites typically involve geotechnical engineers and geologists.

Other sustainable development issues such as renewable energy, greenhouse gas control, and air pollution control also require civil engineers in the design and construction of various plants and energy farms.

## 1.14 Contemporary Issues: The IT Revolution

The Information Technology revolution since the 1990s, with the advent of personal computer in the early 1980s and the internet in the late 1980s, touches every aspect of modern living today. It is a fact, albeit less well known, that the oldest engineering field, civil engineering, has been a heavy user and promoter of high-speed computing. As described in Chapter 5, by the 1960s the stage was set for solving structural design problems using digital computers. The demand for large computer memory for structural problems that contained thousands of unknowns hampered the progress. In the meantime complex interactive graphical display software was developed by civil engineers, architects, medical researchers, and others for applications in their fields. The advent of personal computers, laptops, and tablets and ever-decreasing cost of computer memory finally made it easy for civil engineering–related complex computing. Today, not only structural engineers perform high speed computing on a daily basis, but civil engineers in virtually every other technical area of civil engineering also use computers professionally. New software applications, or new versions of existing applications, are introduced every year. IT literacy becomes a pre-requisite for civil engineers.

The impact on civil engineering from the IT revolution goes beyond the direct use of computers. Civil engineers are heavy users of instruments, from surveying equipment to sensing equipment for earthquake motions and flow monitoring. The surveying equipment today is much easier to use and with much less error than that of a generation ago because of digital computing, not to mention the well-known applications of the Global Positioning System (GPS). In general a sensor picks up a signal, the signal is "conditioned" to filter out "noises" and then it is recorded. The IT revolution greatly simplifies the conditioning and recording because digital computers are so fast and inexpensive and the signal processing can be accomplished by miniature micro-processors imbedded in the instrument packages. Furthermore, the use of wireless digital signal transmission allows distant monitoring. This potential is there to monitor the "health" of structures to avoid any impending danger of failure.

## 1.15 Attributes of a Civil Engineer

In order to carry out the engineering work outlined above, a civil engineer is likely to possess certain attributes, in addition to the abilities and skills acquired through the educational process of a Bachelor of Science in Civil Engineering (BSCE) degree program described in Chapter 3 and Chapter 4. These attributes are listed here.

**Analytical and Organized**: The application of the vast amount of knowledge in civil engineering to solving real-life problems requires an analytical approach and an organized mind. Civil engineers are generally very organized in their daily lives as well.

**Bold in Conception and Careful in Details**: Seeking solutions to real-life problems requires a bold conceptual design. Every new project is different and requires the exploration of different solution outlines. In carrying out the details of the design, the civil engineer must be careful in every step to ensure safety and accuracy.

**Creative but Conservative**: Finding new and economical solutions to civil engineering problems requires a creative mind that thinks beyond the accepted and regular practices, but the reliability and safety of the created product requires a conservative assessment in every aspect, because for every civil engineering product, failure is not an option.

**Dependable and Trustworthy**: Civil engineers depend on each other in teamwork. Everyone is entrusted to produce reliable and accurate work. Civil engineering projects are all time-sensitive. On-time delivery is highly valued and even financially rewarded as may be written in a contract.

**Ethical and Honest**: The impact of a civil engineering project is usually wide ranged and concerns the public interest because it entails the creation of a new built environment, large or small. A civil engineer must be ethical in practice and honest in character in order to earn the public trust. In Chapter 6, engineering ethics issues are described.

**Forthright but Personable**: Civil engineers need to communicate with others effectively. Very rarely does a civil engineer work alone. The teamwork necessitated by the nature of the engineering work requires a civil engineer to possess excellent interpersonal skills to be able to fit in and work well with others.

**Passionate About Work**: The authors never met a civil engineer who is not passionate and proud of what they do. Because civil engineer products are everywhere and visible, it is possible to see a civil engineer pointing to a building, a bridge, a river, or other structures and proudly announcing his/her contribution.

## 1.16 What It Takes

To practice in any one or a combination of these civil engineering technical areas, one must have (a) the fundamental education, institutionalized as a Bachelor of Science in Civil Engineering (BSCE) degree, (b) passed the Fundamentals of Engineering (FE) examination, (c) the internship experience under the supervision of a licensed professional engineer for four years after the BSCE degree, and (d) the license to practice civil engineering, regulated by each state through an examination, which can be taken only after (b) and (c). Details of the requirements for (c) and (d) are described in Chapter 7. But the very first step for anyone admitted to a BSCE degree program is to complete all requirements of the degree program and get the degree. That requires the right commitment, attitude, and learning skills. Chapter 2 is devoted to exactly the kind of knowledge and skills that are necessary for obtaining a BSCE degree in a reasonably short period of time.

## Additional Reading Recommendations

1. At www.asce.org you will find the technical groups and institutes within ASCE. A perusal of these contents will give you a sense of the scope of civil engineering.

2. At the ASCE website, you can download a free copy of **Civil Engineering** Body **of Knowledge for the 21st Century**, 2nd Edition, 2008, (1.1 MB). The first one third of the book gives the type of civil engineering knowledge needed in the 21st century.
3. *Studying Engineering, A Road Map to a Rewarding Career*, Raymond B. Landis, 4th edition, Discovery Press (www.discovery-press.com), 2013. Chapter 2 of his book contains descriptions of all engineering disciplines.

## Acknowledgments

The authors are grateful for the contributions/suggestions made by Professors Clark Liu (University of Hawaii), Eugene Tseng (University of California, Los Angeles), and Yue-hwa Yu (National Taiwan University) to part of the content of this chapter.

## Assignments

1. Find photos of at least three different construction machines from the internet among the construction machines mentioned in this chapter. Write a short description of the functions of each machine and cite the sources and copy the photos to your report. The report is to be contained in one page.
2. Write a one-page report to describe a public civil engineering project currently underway or soon to be launched in your state, including the purpose, benefit, cost, and schedule of completion. Describe the technical areas of civil engineering involved in the project.
3. Read the Executive Summary of ASCE's 2013 report Card for America's Infrastructure and your state's success stories described in the ASCE Report (http://www.infrastructurereportcard.org/a/#p/overview/executive-summary). Write a one-page report to describe one of the success stories and update its progress/status now.
4. Select one from the five choices. Earthquake engineering is part of (a) structural engineering, (b) geotechnical engineering, (c) environmental engineering, (d) transportation engineering, or (e) geomatics.
5. Ocean engineering is part of (a) structural engineering, (b) geotechnical engineering, (c) environmental engineering, (d) transportation engineering, or (e) Water Resources Engineering.
6. A landslide occurred near a populous town. Who among the following civil engineering specialists is most likely to be called upon to investigate? Select one from the five answers: a (a) structural engineer, (b) geotechnical engineer, (c) environmental engineer, (d) transportation engineer, or (e) surveyor.
7. When the curvature of the earth's surface cannot be neglected, you must apply (a) plane survey, (b) land survey, (c) geodetic survey, (d) aerial photography, or (e) all of the above, in doing a survey.
8. Which of the following civil engineering technical areas is the knowledge of chemistry and biology most applied? Answer: (a) structural engineering, (b) geotechnical engineering, (c) environmental engineering, (d) transportation engineering, or (e) water resources engineering.
9. Waste water treatment is part of the (a) structural engineering, (b) water resources engineering, (c) environmental engineering, (d) transportation engineering, or (e) geomatics.

10. Drinking water treatment is part of (a) structural engineering, (b) geotechnical engineering, (c) environmental engineering, (d) transportation engineering, or (e) geomatics.
11. The mayor of a large city is forming a committee to study the feasibility of building a new stadium to host a home basketball team. Whom of the following civil engineers are likely to be included: (a) structural engineer, (b) geotechnical engineer, (c) environmental engineer, (d) transportation engineer, or (e) all of the above.
12. Transforming design details on paper into physical reality is the work of (a) structural engineering, (b) geotechnical engineering, (c) environmental engineering, (d) construction engineering, or (e) surveying engineering.
13. The LEED certificate is for (a) green energy, (b) solid waste disposal, (c) water conservation, (d) green building, or (e) nuclear waste disposal.
14. According to the 2013 ASCE Report Card, the cost of renewal of America's civil infrastructure was estimated at (a) 1.2 trillion, (b) 1.6 trillion, (c) 2.2 trillion, (d) 3.0 trillion, or (e) 3.2 trillion.
15. A sustainable development satisfies the (a) social, (b) economical, (c) environmental, (d) economical and environmental, or (e) social, economical, and environmental concerns.

# Chapter 2

# How to Succeed

## 2.1 Overview

You have successfully graduated from high school and have been admitted to a college program to pursue a Bachelor of Science degree in Civil Engineering (BSCE). To successfully complete such a program requires your commitment, a positive attitude, and new learning skills that are different from what you were required to have to be successful in high school.

A BSCE degree program is defined by its curriculum, which will be described in Chapter 3. For the time being, it suffices to say that doing well in every course you are taking is a prerequisite to the eventual success in earning this degree. Your curriculum is different from your high school curriculum in one major aspect: it is full of prerequisites, courses that must be passed before taking on courses following them. These prerequisites are not set up for the sake of putting up roadblocks; your mastering their contents is actually necessary for you to succeed in courses downstream. Thus failure is really not an option. Actually you do want to succeed in any college courses, prerequisites or not.

It is logical then to set your immediate goal: excel in the courses you are taking. The suggestions contained in this chapter are designed to help you achieve your immediate goal as well as overall success in your college career.

## 2.2 Commitment Is Key

One of the authors of this book once served in a university with a predominance of engineering students. The university conducted a study on factors affecting freshman retention, i.e., the percentage of students returning after one year. At the beginning of the freshman year, students were asked to fill in a survey on the degree of commitment to the program they were in. While the factor of commitment is subjectively determined by students, the university also included other more objectively measurable factors such as math preparation, science preparation, family financial background, etc. After one year, the retention of each student was correlated with all factors included in the study. It turned out the degree of commitment correlated the most with freshman retention. It was a more reliable predictor for retention than math preparedness, which was considered the most important to engineering student retention according to common wisdom.

What this study revealed was the power of personal will. If one is determined to succeed, then there is no obstacle one cannot overcome. To overcome obstacles takes commitment as well as learning skills. With commitment there comes the will to learn skills that help overcome any obstacle. These skills are covered below. For most students, however, to succeed does not always mean to overcome obstacles. Rather it means not to take any courses lightly.

## 2.3 Taking Every Course Seriously

When one of the authors was a freshman he witnessed an unbelievable event: one of his high school classmates was flunked out at the end of the freshman year. The classmate was an excellent student in high school. At the time the university had a rule: anyone who failed two-thirds of the courses taken was automatically dismissed. He virtually failed every course he had taken, not just math and science courses but also general education courses. What had happened? Well he was observed to cruise around campus in his new scooter with a new friend every week riding on the back. He hardly studied; was too busy enjoying the newfound freedom as a college freshman.

What one learns from this story is one can fail any course in college. There is no "easy" course in college. A common fallacy for civil engineering students is to believe only math, sciences, and engineering courses are "hard" and everything else is "easy." The authors had studied freshman retention by examining the records student by student and found there are no easy courses no one ever fails. Students who failed freshman courses typically failed in one or more math and sciences courses AND one or more general education courses.

One should approach every course with the understanding that one can fail and should take it seriously. But how? First of all, you should not "fight" the course. We often heard students complain: oh, the course was boring, I am not interested. Well the course is put on your curriculum for a reason. Please do not second-guess the wisdom of your faculty who designed your curriculum every time you feel you are not interested in the subject matter of a course. Practically speaking, you just need to excel in the course, interested or not, and the first step in that direction is to know the requirements of the course.

## 2.4 Knowing the Course Requirements

Instructors are required to make it clear to students what the course requirements are: textbook, homework (if any), quizzes/tests/exams, term project (if any). These are included in the course syllabus, which also includes a course outline. The course outline gives you a general idea on the scope of the topics covered.

If a textbook is required, that means the book needs to be read and studied frequently enough that one should own it. For math, sciences, and engineering courses, owning the right edition is important because often homework problems are assigned from the book and different editions usually have different homework problems. It is advisable to consult the instructor before buying an older edition. We found students doing the wrong homework problems because they were using the older edition.

You need to have access to the book as early as possible. Even before the first class meeting you can read the preface of the book and go over the table of contents to get a general idea of the scope and subjects covered in the book.

Knowing the test frequency helps you manage your time more effectively for before-test review when all the courses are considered. Time management is of paramount importance and is covered at the end of this chapter.

## 2.5 Attending Classes Religiously

Class meetings are important for the following reasons:

1. They are the only venue from which students can learn from instructors and sometimes interact with them as well for an extended time and on a regular basis.
2. They are the only venue in which students can learn by listening, watching, thinking, and writing during the meeting time.
3. They are the only venue students can listen to and watch fellow students when they ask questions, thereby know the key matters of concern to others.
4. They are the only venue where additional insights on the lecture subject can be learned from the instructor and other students (if active interaction is prevalent).
5. They are the only venue in which stimuli from lecture and discussions can lead to in-depth understanding of the subject.

To reap the above benefits, you need be fully engaged in terms of concentration and participation during class meetings. One can start by attending classes on time and be there a few minutes before a class starts and be seated within good hearing range and good sight of the lecturer's moving space. Many instructors give an overview of the whole lecture at the very beginning. Missing the first few minutes of the lecture is to miss the direction of the whole lecture.

Missing classes seems to be a favorite expression of the newfound freedom of college students. Stories are abundant from successful people who took pride in skipping classes for other favorite activities. A famous example, perhaps, comes from the 42nd president Bill Clinton. When he was studying at Georgetown University, he routinely skipped classes for political activities. When exam times came he would cram on notes from friends and, unbelievably, always would pass. Well, he was not a civil engineering student. The subjects of math, sciences, and engineering are difficult, if not impossible, to digest and master in a short period of time (cramming before exams simply does not work). And, most of us are not endowed with the special ability of Clinton. Consider this: during one of his state of the union addresses, the teleprompter was feeding him an old version of his address that he had already completely revised. He went on to deliver his new address from memory for about ten minutes before his staff discovered the mistake and fed the right version into the teleprompter. Extraordinary people may be entitled to exceptions. You would benefit, however, by not assuming you are an extraordinary civil engineering student but by trying to earn the grades with proven practices.

Another reason we hear all the time from students skipping some classes: I don't like the instructor. Well, you don't have to like the instructor to excel in the course. You don't want to fail because you don't like the instructor either. Missing classes is one of the major reasons students fail a course.

There is one more practical reason that you should not miss any classes. The time needed to catch up with and make up the subject material of a missed class meeting is much more than the time saved by missing the meeting. Furthermore, most students skip a meeting to do something else, not to study the subject material on his/her own during the meeting time. That means additional time is needed to study the subject material and that may very well lead to a downward cycle of digging deeper into a hole of needing even more time to study.

## 2.6 Preparing for Class Meetings

Besides attending classes religiously and on time, another good practice is to read the textbook and review notes from the last meeting before the class. We regularly ask students how that practice benefits them. These are the benefits they told us:

1. The lecture is much easier to follow, because they already have an idea on what to expect, although only vaguely.
2. It is easier to interact with the instructor by asking questions that have emerged from pre-class reading.
3. It is much easier to remember the subject material because the lecture reinforces the pre-class reading.
4. It takes only a fraction of the time of the meeting to prepare for the meeting.

They also told us that they often used time spent on buses or between classes to read through textbooks quickly. They picked up things that were not fully understood with the quick read and paid special attention when the topics were covered during the class meeting, and if necessary, asked questions.

## 2.7 Making the Most of Class Meetings

Students learn in different ways and instructors lecture in different ways. No instructor can lecture the way every student appreciates. Most instructors try to lecture the way most students in the class appreciate. That means sometimes a student does not appreciate the way an instructor lectures and may choose to skip the class. That is a mistake because the person who suffers the most from the consequences is the student already described.

From your perspective, it can only be a lucky coincidence that all your instructors lecture in the way you appreciate. You will benefit by adjusting to the way the lecture is delivered and not to be judgmental about the instructor. No matter how the lecture is conducted, there are skills you can use to gain the most from the class meetings:

1. Concentrate on what is being presented, not what one expects to be presented. In other words, make oneself an empty vessel so that one can absorb whatever is being poured into it.
2. Concentrate on the concepts behind any new terminology and grasp and remember the definition of each new term.
3. Concentrate on the new tools being introduced and the different ways the tools are applied to solving problems.
4. Take notes on what is in addition to the textbook materials. This is assuming you have previewed the subject being covered in the textbook.
5. Practice interactive learning by asking questions and answering any questions the instructor asks.

Taking notes is a good practice but notes must be in clear handwriting so that they are easy to review. Using laptops/tablets for note-taking is excellent especially if you can also draw figures with them.

## 2.8 Practicing Post-Class Review

Learning is a process of continuing to digest and absorb new knowledge and skills. Reviewing the textbook and notes after a class serves to solidify what you have learned in class and discover what is still not clear to you despite the class meeting.

When reviewing textbooks, especially for math, sciences, and engineering courses, one pays special attention not only to pertinent mathematical formulas and equations but also to the **example problems**, even when they are already covered in the class.

Usually there are multiple example problems. One should go over the solution process quickly; follow the analytical flow and logic of the solution and appreciate the path from the given known factors to the solution of the unknown factors. One should also analyze the difference in consecutive example problems to see the change of formulation of the problem in terms of the lineup of known factors versus the unknown factors. This is the way to thoroughly understand a subject. Only through understanding will you remember the solution methods and their natural logic flow.

When you have reviewed the textbooks and notes, you are ready to tackle the homework problems.

## 2.9 Doing Homework Assignments Diligently

Most instructors of math and science and civil engineering courses assign homework problems. Doing homework is a critical part of learning. In attending classroom lectures and from reading textbooks, you learn the concept of a subject and the tools in problem solving. The example problems in the textbook demonstrate the use of tools for problem solving. Students are tempted to feel that they have already learned everything there is to learn about the subject. The key question, however, is whether or not they can solve problems using the concept and tools just learned.

Many years ago a student got an F in Statics, a required course in civil engineering, from one of the authors. The student came to complain and said "I feel I understand all the subjects and can teach the course, but you gave me an F." Well, he failed to mention that he skipped all the homework assignments and was thus without the benefit of testing his problem-solving skills. An ability not tested is an ability not acquired. Homework also plays the role of reinforcing the memory of what has just been learned. You not only understand but now you also remember.

There are three **levels of difficulty** in homework problems. The easiest are those problems similar or even identical to the example problems. Doing those problems serves the purpose of an exercise or a drill. The second level of difficulty is for those problems that are variations of example problems such as changing the role of variables from unknown to known or vice versa. The highest level of difficulty is for problems involving synthesis of two or more concepts or using two or more tools. Those problems often appear as out of the blue and require careful analysis of what is given and what is to be found.

It is important to develop the ability to ascertain the correctness of one's answer, not by checking to see if it is identical to what is given at the end of the book, but by verifying the steps taken in arriving at the answer. That involves checking the solution method itself and the numerical computation. The ability to **double check** one's own work is at least as important as the ability to do the problem. After all in real-life engineering problems, there won't be answers at the end of a book and you are the only one who can ascertain the correctness of your results.

Time spent on doing homework is time well spent in the learning process and should be budgeted generously in the time-management section later in the chapter.

For classes in which homework assignments are graded and will be part of the overall term grade, one question often comes up: can a student participate in **group learning** by doing homework together in a group? The best policy is to ask for guidance from the instructor. If the instructor explicitly forbids doing homework together, then clearly students should follow the guidance. Even if the instructor allows it, students should understand the limit of group learning and the purpose of doing homework as delineated above. The proper way to do homework together is to discuss the sticking points of the homework problems and how you overcome these sticking points. You should still work out the solution on your own and earn the grade on your own. After all one the purpose of homework is to test **your** ability to solve problems.

For essay-type homework problems, you would benefit to learn to present the solution in a logical and clean manner whether they are graded or not. The ability to **present solutions well** can be learned by practicing it constantly. There are two ways of presenting a solution well: Do it first on scrap paper then copy a clean version or to think through in the mind and organize it first and write down a clean version. The former is often necessary for complicated problems involving several steps and the latter is doable for simpler problems. It is a good practice to always try to mentally work out a clear path to a solution before putting anything down on the answer sheet.

This brings up the obvious question: what is the **best strategy for problem solving**? In the context of solving homework problems and taking tests, we suggest the following steps:

1. Read the problem statement carefully to find out what you are asked to determine.
2. Find out what the variables/factors are that are given and known.
3. Figure out the path connecting the known to the unknown.
4. Formulate the solution in terms of mathematical expressions.
5. Carry out the mathematical manipulations toward the solution for the unknown.
6. Review the formulation and then the mathematics.
7. If possible, check the answer to see if it fits the given factors.

Sometimes one simply gets stuck in the process and cannot move forward toward a solution. We always advise, in case of homework problems, to **seek help** if you have already spent more than 20 to 30 minutes on a problem and cannot move forward despite all efforts. There are several possibilities that your university offers:

1. Teaching assistants: Many universities hire graduate students as teaching assistants who should be able to assist undergraduate students in solving homework problems.
2. Learning/tutoring center: Many universities or colleges organize their learning centers to assist students in problem solving. Tutors are graduate students or upper-class undergraduate students who have already taken most of the courses for which some students seek help.
3. Instructors: Most universities have explicit policies on faculty office hours. Students should be able to go to the instructor during those posted office hours or make appointments outside of the office hours if the instructor encourages it.

When seeking help in solving homework problems, one should never expect a tutor or an instructor give a solution line by line. When helping students we always ask students to explain first what the sticking points are. We then ask leading questions and encourage students to think along step by step. When the sticking points are clarified and you are clear on the path to the solution, you can do the rest successfully on your own.

We now offer a relatively simple example problem, which the authors were taught in their elementary school days. We find it useful to illustrate the problem-solving process without anything other than arithmetic and simple reasoning.

**Problem.** Numerous rabbits and chickens are confined in a big cage. There are 50 heads and 140 legs. How many rabbits and how many chickens are in the cage? Solve this problem without using algebraic equations.

**Solution.** We shall go through all the steps outlined above one by one:

1. *Read the problem statement carefully to find out what you are asked to determine.*
   You are asked to find the number of rabbits and the number of chickens.
2. *Find out what are the variables/factors that are given and known.*
   You know the number of heads and legs, **50** and **140** respectively. **You also know** that each rabbit has **4** legs and each chicken has **2** legs.
3. *Figure out the path connecting the known to the unknown.*
   The **Key** is the difference in the number of legs of the two animals.
   If the cage contained only chickens (2 legs), there would be a total of only 50×2 = 100 legs.
   The 40 additional legs (140 – 100) are due to the 2 additional legs for each rabbit in the cage.
   Dividing the 40 by 2 gives the number of rabbits.
4. *Formulate the solution in terms of mathematical expressions.*
5. *Carry out the mathematical manipulations toward the solution for the unknown.*
   **The difference in the number of legs between one rabbit and one chicken: 4–2** = 2
   **If rabbits had two legs each, the number of legs would be** 50 × 2 = 100
   **The additional legs due to rabbits having 2 more legs:** 140–100 = 40
   **The number of rabbits must be** 40/2 = 20
   **The number of chickens must be** 50–20 = 30
6. *Review the formulation and then the mathematics.*
   The above computation is accurate.

7. If possible, check the answer to see if it fits the given factors.
   20 + 30 = **50** Check!
   20 × 4 + 30 × 2 = **140** Check!!

Now try to solve this problem focusing on the fact that a chicken has **2 fewer** legs instead of a rabbit has **two more** legs. This is one of the assignments at the end.

## 2.10 Creating a Course Portfolio

It is beneficial to have a course portfolio in the form of a three-ring binder so that course syllabus, notes, handouts, homework solutions, graded exams, and solutions can be organized by category and within each category in chronological order. The portfolio makes review easy, be it for exam preparation, for doing homework, or for preparation several terms later for the Fundamentals of Engineering exam described in the next chapter.

For each course there should be one course portfolio. We observed that many students like to have a large notebook for every course taken in the same term, but that is not a course portfolio. A portfolio is a much broader collection of important materials for a course, as just described. Learning to organize a portfolio is an important experience for a civil engineering student. An added benefit for having course portfolios is to show off your organizational skills and achievements in any particular course to your potential employers when the time comes for job interviews.

Many BSCE programs begin to ask students to create course portfolios for ABET review. ABET is the organization that reviews and accredits engineering and technology programs. ABET formerly stands for Accreditation Board for Engineering and Technology but now its formal name is simply ABET Inc. ABET-related topics are covered in the next chapter.

## 2.11 Interacting with Instructors

It is not unusual that students do not know the name, the office location, the email address, or telephone number of instructors. Just ask yourself and you would agree. Most universities today encourage faculty–student interaction beyond classroom meetings. Most faculties encourage students to visit them if students have questions. From students' perspectives they do need to find out first if communication with the instructor is welcome and how best to reach them. Some instructors encourage email communications but discourage office visits. As long as interaction is encouraged, one should fully utilize this particular resource. Benefits to students include:

1. Instructors are most equipped to answer questions from students in homework and course content and may be willing to provide study guidance beyond simply solving problems.
2. Instructors may be able to provide career advisement in general.
3. Instructors may be able to provide internship and summer job recommendations or even job recommendations after graduation.

In seeking recommendations from instructors for jobs, scholarships, or awards, you are encouraged to choose those who are most likely to give credible, strongly positive recommendations. They must have direct knowledge of your academic or extracurricular activities or performances. Otherwise a recommendation letter they write will have no effect or even negative effects on your opportunities. One simple rule is to seek out instructors with whom you have excellent grades or working relationships, such as assisting in laboratory work or research work or even grading.

## 2.12 Studying with Classmates

It is not uncommon, especially among those who commute to campus, that students hardly know any classmates. For one thing, students go from class to class usually in a rush and hardly have time to talk to each other. For another, they may not have many classes in common. As a result students become loners as far as taking courses is concerned: they go to a class alone and study alone. If you are used to studying alone and still excel in course work, so be it.

Many educators today believe **group study** is beneficial to most students. The study group typically of three to five students becomes a network. They help each other if anyone missed a class for good reason or simply missed some key points during a class meeting. The sense of belonging to a group is by itself beneficial, according to educators. The group is a source of support and a place to learn interpersonal communication skills as well. A group for a course may be formed naturally among students who know each other. Or it is formed with a leader who organizes the group. Either way, the first step is to know each other in a class. Knowing someone's first name makes starting a conversation easy. Many instructors encourage students to know each other by asking students to introduce themselves during the first meetings.

The best place for group study is either a studying space designated by your program or in the group study rooms set aside by the university library.

## 2.13 Using the University Library

In the age of computers and internet, the library is often ignored by students. But many university libraries have caught up with the internet age and offer online searches and book ordering. Another important service to students is providing study space. In addition to the traditional individual studying space, such as tables and chairs or even individually reserved study carrels, many libraries provide group study rooms that are enclosed and relatively sound-proof with large glass doors or windows. These rooms are perfect for small group discussions.

Other resources of a university library include new media materials such as DVDs and CDs. Interlibrary services allow students to order books or journal articles from almost any other library.

So far we have suggested several things for you to do, but we know many students have complained to us: we don't have time to do all these things. But you do, if you know how to manage your time.

## 2.14 Time Management: Exam Preparation

Common wisdom dictates that one shall never walk into an exam unprepared or under prepared. There is no chance that you can simply guess a solution right. Solutions must come from what you have learned. Preparing for exams is much easier and less time consuming if you keep up with the progress of a course and review and do homework assignments promptly. Still a plan for preparation for an exam is always beneficial. Shown below is a simple worksheet for exam preparation.

The studying time needed is to be matched by the available time total. If you keep up with the course progress on weekly basis, then the time needed before an exam is expected to be much lower than otherwise. The number of days before the exam day to be included in the above planner can be adjusted accordingly.

Table 2.1 An Exam Planner

**Exam Planner for** ______________

**Exam Date**: ______________

**Exam Coverage**:

________ Chapters of Textbook, ________ Pages
________ Pages of notes
________ Example problems
________ Homework problems
________ Other reference materials

**Study hours needed before the exam**:

____________for textbook
____________for notes
____________for example problems
____________for homework problems
____________for other reference materials
____________for general overview

**Study hours needed total:** ____________hours

**Devoted Study Time**—Exam Date minus 1: ____________hours
**Devoted Study Time**—Exam Date minus 2: ____________hours
**Devoted Study Time**—Exam Date minus 3: ____________hours
**Devoted Study Time**—Exam Date minus 4: ____________hours
**Devoted Study Time**—Exam Date minus 5: ____________hours
**Devoted Study Time**—Exam Date minus 6: ____________hours
**Devoted Study Time**—Exam Date minus 7: ____________hours

**Available Time Total:** ____________hours

## 2.15 Time Management: A Weekly Planner for Success

**Your need for study time.** The knowledge embedded in mathematics and basic sciences courses and all the engineering sciences and civil engineering courses are cumulative in nature. Within each course the knowledge learned in this week may be needed to learn what is going to be covered in the next week and the weeks to follow. Another important feature of these courses is the new concepts and new tools introduced weekly. The new knowledge is to be understood thoroughly and remembered, and new tools need to be practiced and mastered. That means:

(a) Homework assignments must be taken seriously and worked on as soon as possible but not before the textbook and notes taken during classes are thoroughly reviewed.
(b) Exams must be thoroughly prepared well in advance of the exam date by reviewing the textbook, notes, and homework problems worked on and solved.

Every student must develop a sense of how much time is needed to carry out the above tasks for a particular course. In the book *Study Engineering* by Raymond Landis (see suggested reading at the end), he cited a **60-hour rule** credited to Dean Tom Mullinazzi. The 60-hour rule is based on two assumptions. First it is assumed that a student can spend 60 hours a week over the period of a term on academic study, working a paying job, and commuting. It also assumes that a student studies two hours for every hour spent in the classroom. This rule provides a useful guidance on how many credit-hours a student can take without overburdening oneself.

As an example, let us consider someone spends 15 hours on a job and a negligible amount of time commuting. The number of hours left for studying is 60–15 = 45. For each credit-hour taken one needs 1 + 2 = 3 hours of studying. Thus the maximum number of credit-hours to take is 45/3 = 15.

When you apply this rule, you can adjust both assumptions based on your own situation. We learned in the past that some students committed 70 hours or more a week instead of 60 hours. You may need more than three hours for some courses and fewer for some other courses.

Now you work on a weekly planner as an important tool for effective time management. You review the actual time spent on each of the following activities in a typical week.

**Sleep (SP)**: Everyone sleeps every day of the week but may not sleep the same number of hours from day to day. A day-by-day review and summary will reveal how many hours are spent each day of a week. Medical experts advise that for an average adult, at least seven hours of sleep a day is needed, with eight hours being the preferred amount.

**Work (WK):** Many students do work part time or even full time. A daily summary of the hours spent on work gives one the total number of hours spent in a week.

**Attending Classes (CL):** Attending classes is a privilege and an obligation. In a semester-based system, for every credit-unit taken there is close to one hour in class meeting time.

**Eating (ET):** Eating is a daily necessity. Reviewing your time spent on eating does not imply you should eat fast in order to save time; medical experts would not approve eating fast. It is merely necessary to account for the time spent on eating.

**Commuting (CM):** Time spent on moving from one place to another varies greatly depending on whether one lives on campus or off campus in a faraway location. The obvious advantage of living on campus or near campus is the time saved from commuting.

**Studying (ST):** Studying can be achieved with the combination of long periods of time and short periods of time. If there is one hour between classes, 30 to 50 minutes of it can be effectively used for studying by reviewing what transpired in the last class or previewing what is going to be covered in the next class.

**Relaxation and the Rest (RR):** This is a catch-all category. Any other activities that are not accounted for by the above six categories are lumped into this category. Many students spend a significant amount of time every week on recreational activities such as web surfing, listening to music, playing computer games, or talking to friends either face to face or by text messaging and cell phone. Recreational and relaxation activities are necessary in anyone's life.

The easiest way of reviewing how the 168 hours of a week is being spent is to use a simple table as shown below **in Table 2.2**. Using the seven notations: SP, WK, CL, ET, CM, ST, RR, one fills in the half-hour blocks with one of the seven notations day by day. The weekly sum of hours spent on each of the seven categories is then added up and put in **Table 2.3 Weekly Activities Summary**. These are the numbers for the actual situation. We observed that many students were surprised at the numbers and decided to make some changes.

The key question to ask is if the time spent on studying is sufficient to ensure the success in every course you are taking. One useful guide is the average number of hours needed in studying mentioned above: two hours for every one hour in the classroom. Studying time for every week should be maintained more or less constant from one week to another. Upon an honest review and assessment of the weekly activities, many students will make changes, often adding more time to studying.

Every hour added to studying must come from the hours reduced from other activities. It is not advisable to reduce sleep time unless one is spending more than eight hours a day and one's doctor approves of the reduction. In most cases, hours can be relocated to studying from the RR category relatively painlessly. You can now use the same tables as your new weekly planner.

Once you decide to make specific adjustments, you shall commit to the change and review the effects of the adjustment from time to time.

There are more learning skills and useful attitudes for success contained in the open literature. Some are given below.

## Additional Reading Recommendations

1. *Studying Engineering: A Road Map to a Rewarding Career*, Raymond B. Landis, 4th edition, Discovery Press (www.discovery-press.com), 2013. This book contains numerous learning skills for college study and a variety of winning attitudes and behaviors for a rewarding career.
2. *Getting Things Done: The Art of Stress-Free Productivity*, David Allen, Penguin Books, 2001. This book contains numerous suggestions on how to use your time to achieve your goals.

## Assignments

1. Examine your weekly activities and need for change by completing the follow steps:

   (a) Fill in the Typical Weekly Activities table to record the seven categories of activities for a typical week. Use a computer-based spreadsheet software.

Table 2.2 Typical Weekly Activities

| 24-hour day | Sun. | Mon. | Tue. | Wed. | Thu. | Fri. | Sat. |
|---|---|---|---|---|---|---|---|
| 0-5 | | | | | | | |
| 5-5:30 | | | | | | | |
| 5:30-6 | | | | | | | |
| 6-6:30 | | | | | | | |
| 6:30-7 | | | | | | | |
| 7-7:30 | | | | | | | |
| 7:30-8 | | | | | | | |
| 8-8:30 | | | | | | | |
| 8:30-9 | | | | | | | |
| 9-9:30 | | | | | | | |
| 9:30-10 | | | | | | | |
| 10-10:30 | | | | | | | |
| 10:30-11 | | | | | | | |
| 11-11:30 | | | | | | | |
| 11:30-12 | | | | | | | |
| 12-12:30 | | | | | | | |
| 12:30-13 | | | | | | | |
| 13-13:30 | | | | | | | |
| 13:30-14 | | | | | | | |
| 14-14:30 | | | | | | | |
| 14:30-15 | | | | | | | |
| 15-15:30 | | | | | | | |
| 15:30-16 | | | | | | | |
| 16-16:30 | | | | | | | |
| 16:30-17 | | | | | | | |
| 17-17:30 | | | | | | | |
| 17:30-18 | | | | | | | |
| 18-18:30 | | | | | | | |
| 18:30-19 | | | | | | | |
| 19-19:30 | | | | | | | |
| 19:30-20 | | | | | | | |
| 20-20:30 | | | | | | | |
| 20:30-21 | | | | | | | |
| 21-21:30 | | | | | | | |
| 21:30-22 | | | | | | | |
| 22-22:30 | | | | | | | |
| 22:30-23 | | | | | | | |
| 23-23:30 | | | | | | | |
| 23:30-24 | | | | | | | |
| SUM | 24 | 24 | 24 | 24 | 24 | 24 | 24 |
| **SP:** sleep; **WK:** work; **CL:** class; **ET:** eat; **CM:** commute; **ST:** study; **RR:** relaxation & the rest | | | | | | | |

Table 2.3 Weekly Activities Summary

| **Activities** | | **Hours** |
|---|---|---|
| SP | Sleep | |
| WK | Work | |
| CL | Class | |
| ET | Eat | |
| CM | Commute | |
| ST | Studying | |
| RR | Relaxation and the Rest | |
| | | **168** |

(b) Add up the total number of hours spent on each of the seven categories of activities and fill in the Weekly Activities Summary table. Make a copy of the Weekly Activities table and sort for the seven categories (some software does not keep a copy after you sort).

(c) Assess the time spent on studying vs. class time (target ratio 2:1 unless you have your own rule).

(d) Decide on any changes that are necessary and be specific.

(e) Create a new Typical Weekly Activities table and a Weekly Activities Summary table to reflect the specific changes.

2. Numerous rabbits and chickens are confined in a big cage. There are 35 heads and 140 legs. How many rabbits and how many chickens are in the cage? Solve this problem without using algebraic equations. Concentrate on the fact that a chicken has two **fewer** legs than a rabbit.
3. An instructor has a class with 150 students enrolled in it. In a class meeting she noticed that many but not everyone showed up. When she asked the students to self-count by every 3 ( i.e., call out 1,2,3,1,2,3 … ) she observed that there were 2 students left (i.e., the total number of students is 3N+2, N is unknown). She repeated the count for every 5 students, she found 3 were left. Counting every seven students just left 1. What is the total number of students in the classroom? Solve this problem without using algebraic equations.
4. The very first step in problem solving is (a) read the problem statement carefully to find out what you are asked to determine, (b) find out what are the variables/factors that are given and known, (c) figure out the path connecting the known to the unknown, (d) formulate the solution in terms of mathematical expressions, or (e) carry out the mathematical manipulations toward the solution for the unknown.
5. The very last step in problem solving is (a) find out what are the variables/factors that are given and known, (b) figure out the path connecting the known to the unknown, (c) formulate the solution in terms of mathematical expressions, (d) carry out the mathematical manipulations toward the solution for the unknown, or (e) find a way to proceed to check the answer on your own.
6. How does reading the textbook and reviewing notes from last meeting before the class benefit your learning? Select all that apply: (a) The lecture is much easier to follow, (b) It is easier to interact with the instructor by asking questions that have emerged from pre-class reading, (c) It is much easier to

remember the subject material because the lecture reinforces the pre-class reading, (d) It takes only a fraction of the time of the meeting to prepare for the meeting, and (e) It pleases the instructor.

7. You have been working on a homework problem for 30 minutes but are getting nowhere. What should you do? Select the one answer that you should not do: (a) Try to find the solution from a solution manual, (b) work on the problem a few minutes more, (c) seek help from a tutor, (d) seek help from the instructor, or (e) seek help from fellow students.
8. According to the 60-hour rule, what is the maximum number of credit-hours you can take in a semester if you are working 10 hours a week and spending 5 hours per week commuting? (a) 10, (b) 15, (c) 20, (d) 25, or (e) 30.
9. What is a legitimate reason for not concentrating during class time: (a) do not like the course, (b) do not like the instructor, (c) tired, (d) have to work on next class' homework, or (e) none.
10. What are the benefits of interacting with your instructors? Select all that apply: (a) instructors can answer your questions on homework and course content, (b) instructors will give you better grades, (c) instructors may be able to advise on your career plan, (d) instructors may be able to write recommendation letters for you, and (e) in case you need to borrow money, instructors can help you.
11. What is a legitimate excuse for missing a class? (a) need to study for another class, because the test is tomorrow, (b) losing interest in the course, (c) missing just one class is not going to matter, (d) scheduled a job interview in conflict with the class time, or (e) a medical or family emergency.
12. What is the best time to start reviewing for the next test: (a) one week before the test date, (b) one day before the test date, (c) figure out the required study hours and available hours and try to match, (d) when you have no other commitments, or (e) at least four weeks before the test date.
13. Students may fail in what type of courses: (a) math, (b) science, (c) engineering, (d) general education, or (e) all of the above.
14. Doing homework is important. Therefore you should: (a) find a solution manual and study hard, (b) review the textbook, example problems and notes then attempt the homework problems, (c) try to solve homework problems as soon as possible, (d) look at the answer in the back of the textbook to see if there are any clues, or (e) go to a tutoring center to find solutions.
15. In your Time Management exercise you discovered the studying time you allocated every week needs a 10-hour boost. The best place to find the 10 hours is to cut down time spent on (a) sleep, (b) meals, (c) class, (d) work, or (e) relaxation and the rest.

# Chapter 3

# The Civil Engineering Curriculum

## 3.1 Overview

Upon completion of an undergraduate civil engineering education, one receives a Bachelor of Science in Civil Engineering (BSCE) degree. In the United States of America, BSCE and other engineering and technology degree programs are accredited by the ABET Inc., formerly the Accreditation Board for Engineering and Technology, Inc. The purpose of accreditation is to serve the public interest by providing quality assurance of accredited degree programs or institutes. Receiving accreditation is important in higher education because 1) accreditation is the primary means to assure academic quality to students and the public, 2) accreditation allows students to gain access to federal funds (such as the Pell grant) in U.S. (especially the regional and national accreditation of universities and institutes), 3) accreditation facilitates transfer of credit among similar programs, and 4) accreditation provides employers with important information when they make decisions on job applicants and continuing education for their employees. To get a BSCE program accredited, the requirements of ABET are the common base from which the unique features of a particular program are built. As a result, while no two BSCE programs are identical in their curricula, all civil engineering curricula have a common structure. To understand this common structure, you need to understand ABET requirements and the role of ABET in higher education accreditation in America.

In this chapter you will see the ABET requirements for your program as well as your program's curricular structure. Descriptions are given for many civil engineering courses for you to gain an overall sense on the courses you will soon take.

## 3.2 Higher Education Accreditation and ABET

ABET is one of more than 60 specialized and professional accreditors in the United States. Specialized accreditors accredit special programs in law schools, medical schools, engineering and technology schools, and health professional programs. In addition to specialized and professional accreditors, there are regional and national accreditors. The regional accreditors are institutional accreditors who review entire institutes. There are eight regional accreditors covering six geographical regions in U.S. National accreditors are also institutional accreditors that accredit single-purpose institutions, private career institutions, and faith-based colleges and universities. There are eleven national accreditors.

These accreditors are private, nonprofit organizations. To gain the public trust, these accreditors must be "recognized" by either the United States Department of Education or the Council for Higher Education Accreditation (CHEA). The former is a cabinet-level federal government organization and the latter is a private, nonprofit organization. CHEA is governed by a 17-person board of college and university presidents, institutional representatives, and public members. ABET is recognized by CHEA. That means ABET satisfies the following CHEA standards:

TO ADVANCE ACADEMIC QUALITY. To confirm that accrediting organizations have standards that advance academic quality in higher education; that those standards emphasize student achievement and high expectations of teaching and learning, research, and service; and that those standards are developed within the framework of institutional mission.

TO DEMONSTRATE ACCOUNTABILITY. To confirm that accrediting organizations have standards that assure accountability through consistent, clear, and coherent communication to the public and the higher education community about the results of educational efforts. Accountability also includes a commitment by the accrediting organization to involve the public in accreditation decision making.

TO ENCOURAGE, WHERE APPROPRIATE, SCRUTINY AND PLANNING FOR CHANGE AND FOR NEEDED IMPROVEMENT. To confirm that accrediting organizations have standards that encourage institutions to plan, where appropriate, for change and for needed improvement; to develop and sustain activities that anticipate and address needed change; and to stress student achievement.

One continually reassesses the accreditation practice: Accreditors are required to undertake self-scrutiny of their accrediting activities.

ABET is a federation of about 30 professional and technical societies representing engineering, technology, computing, and applied science. It does not accredit universities or colleges (that is the responsibility of regional accreditors) but individual programs. It accredits about 2,500 programs at over 550 colleges and universities in U.S. in 2013. Among the 2,500 programs, about 220 BSCE programs are accredited by ABET.

For over 60 years, the ABET (and its predecessor the Engineers Council for Professional Development) requirements remained structurally the same: it prescribed the curriculum content and other minimum requirements on faculty and facilities. By the end of the twentieth century, these requirements were increasingly considered by the industry as ineffective in ensuring accredited programs were producing graduates that satisfied the needs of the industry. The ABET requirements were considered by engineering educators as too restrictive. By 1997, ABET concluded a ten-year study and put in place a revolutionary approach to the accreditation of engineering programs. The new approach is outcome-oriented and was to be applied to all engineering program reviews by 2000, thus named Engineering Criteria 2000, **EC2000**. It is flexible in its requirements on curriculum, allowing each program to design its own unique features while satisfying some common requirements.

## 3.3 Program Outcomes Requirements of EC2000

ABET has well-defined criteria on the following nine categories: student, program educational objectives, program outcomes, continuing improvement, curriculum, faculty, facilities, support, and program criteria. The first eight are common to all engineering programs. The last is program specific, established by the relevant professional organization associated with ABET. The program outcomes specify the abilities to be acquired by the students by the time they graduate:

(a) an ability to apply knowledge of mathematics, science, and engineering
(b) an ability to design and conduct experiments, as well as to analyze and interpret data
(c) an ability to design a system, component, or process to meet desired needs within realistic constraints such as economic, environmental, social, political, ethical, health and safety, manufacturability, and sustainability
(d) an ability to function on multidisciplinary teams
(e) an ability to identify, formulate, and solve engineering problems
(f) an understanding of professional and ethical responsibility
(g) an ability to communicate effectively
(h) the broad education necessary to understand the impact of engineering solutions in a global, economic, environmental, and societal context
(i) a recognition of the need for, and an ability to engage in life-long learning
(j) a knowledge of contemporary issues
(k) an ability to use the techniques, skills, and modern engineering tools necessary for engineering practice.

These outcomes are commonly known as the a-to-k criteria. In addition ABET encourages each program to add other criteria that are unique to the program. It is beneficial for you to learn what the outcomes of your BSCE program are and how the outcomes are linked to the courses you are required to take. Your advisor can provide you with specific information on this linkage, but you will learn from the following what the general structures of your curriculum are.

## 3.4 Curriculum Requirements of EC2000

ABET specifies the structure and subject areas of curriculum but not the specific courses:

> The curriculum requirements specify subject areas appropriate to engineering but do not prescribe specific courses. The faculty must ensure that the program curriculum devotes adequate attention and time to each component, consistent with the outcomes and objectives of the program and institution. The professional component must include:
> (a) one year of a combination of college-level mathematics and basic sciences (some with experimental experience) appropriate to the discipline

(b) one and one half years of engineering topics, consisting of engineering sciences and engineering design appropriate to the student's field of study. The engineering sciences have their roots in mathematics and basic sciences but carry knowledge further toward creative application. These studies provide a bridge between mathematics and basic sciences on the one hand and engineering practice on the other. Engineering design is the process of devising a system, component, or process to meet desired needs. It is a decision-making process (often iterative), in which the basic sciences, mathematics, and the engineering sciences are applied to convert resources optimally to meet these stated needs.

(c) a general education component that complements the technical content of the curriculum and is consistent with the program and institution objectives.

Students must be prepared for engineering practice through a curriculum culminating in a major design experience based on the knowledge and skills acquired in earlier course work and incorporating appropriate engineering standards and multiple realistic constraints.

The above requirements are for all engineering programs. Each specific engineering program may subject to additional requirements developed by the relevant engineering professional organization.

## 3.5 BSCE Curriculum Requirements

The lead society responsible for the formulation of BSCE curriculum requirements for ABET is the American Society of Civil Engineers (ASCE). The resulting requirements are described in the following single paragraph:

> The program must prepare graduates to apply knowledge of mathematics through differential equations, calculus-based physics, chemistry, and at least one additional area of basic science, consistent with the program educational objectives; apply knowledge of four technical areas appropriate to civil engineering; conduct civil engineering experiments and analyze and interpret the resulting data; design a system, component, or process in more than one civil engineering context; explain basic concepts in management, business, public policy, and leadership; and explain the importance of professional licensure.

## 3.6 BSCE Curriculum Structure

A BSCE curriculum satisfying the general EC2000 requirements (Section 3.4) and the BSCE requirements (Section 3.5) has five elements in its structure: General Education, Science and Mathematics, Engineering Science and Civil Engineering Fundamentals, Civil Engineering Technical Areas, and Capstone Design. While no two BSCE curricula are identical, these five elements are common to all BSCE curricula and add up to 120 to 132 credit units in a semester-based system in most universities. In the following, whenever a course title is used, it is to be understood that the title is meant to be descriptive and your program may use a different title. Sometimes one or two courses may be bundled as a series under a more general title. Your program may offer the opportunity for you to select from an approved list of courses in each of these five elements.

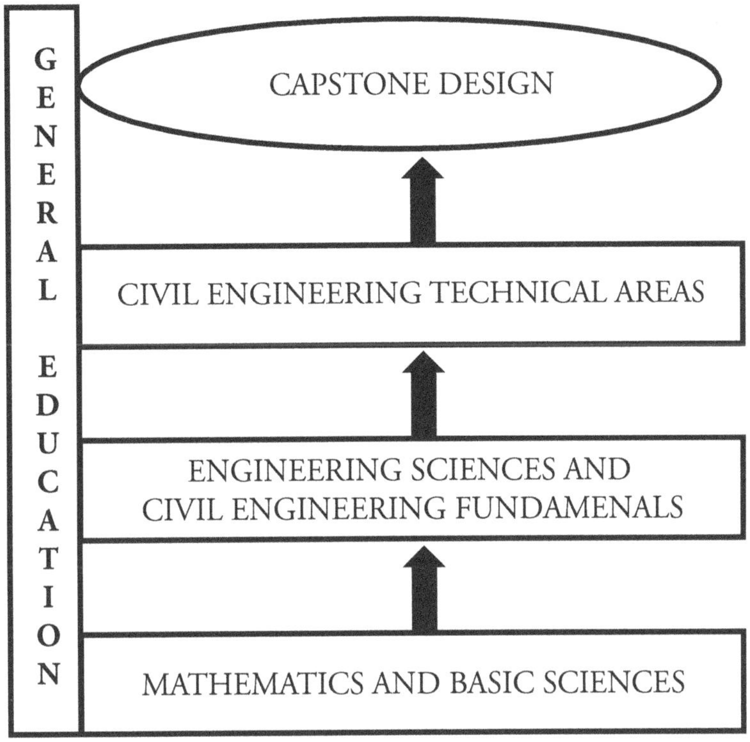


Figure 3.1 Basic Structure of a BSCE Program.

The following diagram summarizes the basic structure of a BSCE curriculum.

**General Education**. Since the 1960s, educators in higher education recognize that a college graduate should have knowledge and skills not only in his/her chosen field of study but also in other areas that are vital to the understanding of a modern society. Each higher learning institute began to select these areas of study to form a core of its curriculum that best defines the institute's educational mission. These areas of study are required for all of its students. Liberal studies students are required to learn basic mathematics and science while science and engineering students are required to learn humanities and social studies. Some state governments also require students in state-supported universities to take history and government courses. These general education courses satisfy EC2000's requirement of "a general education component that complements the technical content of the curriculum and is consistent with the program and institution objectives." Usually you are given the liberty of choosing from a long list of eligible courses from a wide range of disciplines. The general education units add up to roughly one year's worth of study.

**Mathematics and Basic Sciences**. As described in Section 3.5, ABET required for civil engineering students: "knowledge of mathematics through differential equations, calculus-based physics, chemistry, and at least one additional area of basic science, consistent with the program educational objectives."

The mathematics requirement is typically satisfied by one year of basic calculus courses and one year of multivariable calculus, differential equations, and linear algebra courses.

The additional area of science requirement for BSCE is unique among engineering programs and can be satisfied by one single course in biology, geology, or other science courses. The mathematics and science requirements amount to about one year's worth of study.

**Engineering Sciences and Civil Engineering Fundamentals**. The engineering science courses are required for most engineering disciplines. Civil Engineering fundamental courses are required for civil engineering but are not for most other engineering disciplines.

**Engineering science courses** in a civil engineering curriculum typically include: statics, dynamics, thermodynamics, and fluid mechanics. A brief description of each emphasizing its relevance to civil engineering is given below.

**Statics**: Civil engineering structures are subjected to forces that either do not vary in time or do not vary rapidly in time. An example of such forces is the gravitational force commonly known as weight or dead load. According to Newton's First Law, the summation of all forces acting on an object at rest is zero. These forces are said to be in equilibrium. Statics deals with **force and equilibrium** and the tools derived thereof. These tools are to be applied in several subsequent courses in civil engineering.

**Dynamics**: Civil engineering structures are subjected to incidental environmental events such as wind and earthquakes that may cause structures to vibrate. Dynamics deals with motion of objects and the relationship between force and motion. Dynamics forms the basis for the study of the vibration of structures and the motion of fluids in pipelines and open channels. Dynamics and Statics form the two-course series usually called Engineering Mechanics.

**Thermodynamics**: In the study of fluid motion involving the effect of temperature, the application of Statics and Dynamics alone is not sufficient. It requires the knowledge of Thermodynamics, which studies the conversion among different forms of energy: heat, mechanical, electrical, and chemical. Of importance to civil engineering applications is the interrelation of variables such as temperature, pressure, and volume of fluids.

**Fluid Mechanics**: Fluids include gases and liquids. In the context of civil engineering, fluids are water and air. Movement of air is relevant in environmental engineering while movement of water is relevant in both water resources engineering and environmental engineering. Fluid mechanics studies the movement of fluids in open and enclosed environment. It applies the principles and tools developed in Statics and Dynamics to a special medium, fluid, which changes its shape easily.

**Civil Engineering Fundamental Courses** typically include engineering materials (or construction materials), mechanics of materials (mechanics of deformable bodies or strength of materials), soil mechanics, and hydrology.

**Engineering Materials**: Engineering Materials studies the manufacturing and the properties of materials for engineering applications. In the civil engineering context, steel, concrete, and timber are the major construction materials. The main material properties are the mechanical properties such as strength, hardness, toughness, and ductility. The Engineering Materials course for civil engineers often concentrates on steel and concrete and leaves timber to a separate timber design courses.

**Mechanics of Materials**: Mechanics of Materials relates forces acting on structural members (such as prismatic bar, beams, and columns) to the effects on the material of the members. These effects result in changing the shape and size of structural members. Thus, the course is sometimes entitled **Mechanics of Deformable Bodies**. It is also called **Strength of Materials**. In some BSCE programs, the content of this course is included in an Introduction to Solid Mechanics course.

**Soil Mechanics**: Virtually all civil engineering structures are supported by soil or rock foundations. The strength of soil for load bearing is developed in soil mechanics. Unique to the strength of soil is its dependence on the degree of the presence of moisture, or water. The laboratory component of soil mechanics is usually included as a part of the course. In some BSCE programs, the content of this course is included in an Introduction to Geotechnical Engineering course, which may include rock mechanics.

**Hydrology**: Hydrology studies and quantifies the circulation and movement of water. At a local or regional scale, the flow of river water can be predicted under normal or storm circumstances. For example, knowing the rainfall at an upstream region, the level of river water downstream can be computed and proper warning may be issued. Hydrology is the base on which scientific tools are for solving practical water resources problems. In some BSCE programs, the content of this course is included in an introductory Water Resources Engineering course.

In addition to the engineering science and civil engineering fundamental courses just described, some BSCE programs also require one or more of the following: **Electric Circuit** (or Electric Network), which covers electric circuit analysis, transformers, power supplies, etc.; **System Engineering**, which covers the mathematical basis for system analysis of civil engineering systems and its impact on the environment; **Introduction to Computing**, which covers either fundamentals of computer programming or tools of computing for civil engineers; and **Probability and Statistics for Civil Engineers**. All the engineering sciences and civil engineering fundamental courses add up to close to one year's worth of credit units.

**Civil Engineering Technical Areas**. As required by **EC2000**, a BSCE program "must demonstrate that graduates can apply knowledge of four technical areas appropriate to civil engineering." Reflected in a BSCE curriculum for each of the four or more technical areas is at least a two-course series with at least one course in the upper-division level, i.e. junior or senior level. All BSCE programs have structural engineering as one of the four or more technical areas they offer. Beyond structural engineering each BSCE program offers a variety of other technical areas based on the expertise of its faculty. In the following, only the structural engineering area and its three most commonly offered courses are briefly described.

**Structural Engineering Area**: Structural engineering is the earliest developed civil engineering technical specialty area. It has the most undergraduate courses to offer. Beyond the engineering mechanics courses and the mechanics of materials, there are Structural Analysis, Reinforced Concrete Structures Design, and Steel Structures Design.

**Structural Analysis**: Sometimes called Theory of Structures, the Structural Analysis offers further analytical tools for computing the member forces and the deflections of beams, trusses, and frames. The law of materials used in the development of the analytical tools is limited to that of linearly elastic, which means the relationship between stress and strain is proportional, and once the force acting on a structure is removed, the structure members return to its original position. The linearly elastic material law is applicable to most situations when the deflection of a structure is much smaller relative to the dimension of the structure member. More complicated material laws are covered in advanced structural analysis courses.

**Reinforced Concrete Structures Design**: The reinforced concrete structures design course covers beam design, simple slab design, column design, and simple foundation design using concrete and reinforcing steel bars. The design of these structural members follows the design code and specifications developed by professional societies. The design processes include, for a given set of forces acting on a member, the selection of concrete strength and steel reinforcing bar strength, member size determination, and the number and spacing of reinforcing bars in the member.

**Steel Structures Design**: For steel structures, the design deals with tension and compression members, and beams and beam-columns. Unlike concrete structures, which are usually cast in situ to form an integral structure (except for precast members connected in situ), steel structures are formed by connecting steel members with the use of bolts or welding. The design of the connection is unique to steel structures and is governed by design codes and specifications. In most applications, the design of a structural member entails the eventual selection of the member size and strength from a list of manufactured steel sections provided by the steel manufacturing industry.

Other undergraduate courses in structural engineering may include design of timber structures, masonry structures, advanced structural analysis, and pre-stressed concrete structure design.

In addition to structural engineering, many BSCE programs offer geotechnical engineering, water resources engineering, environmental engineering, transportation engineering, and construction engineering as part of the four or more areas required by ABET EC2000.

**Capstone Design Course**. The EC2000 of ABET requires: "Students must be prepared for engineering practice through a curriculum culminating in a major design experience based on the knowledge and skills acquired in earlier course work and incorporating appropriate engineering standards and multiple realistic constraints." Among the EC2000 a-to-k outcomes are two criteria: (c) an ability to design a system, component, or process to meet desired needs within realistic constraints such as economic, environmental, social, political, ethical, health and safety, manufacturability, and sustainability, and (d) an ability to function on multidisciplinary teams. Most BSCE program requires one course or a two-course series of design at the senior level to address these requirements. The teamwork component of the design experience is addressed by requiring the design be completed by a team of students working together. The ability of students to function on a multidisciplinary team is acquired through the capstone design course by requiring each team member to perform different tasks preferably in more than one technical area.

It should be emphasized that each BSCE curriculum is developed over time with historical perspectives and the vision and experience of the program faculty. While the above curriculum structure and courses are described according to the ABET EC2000 framework, it is not implied that ABET creates the BSCE curriculum.

## 3.7 Navigating a BSCE Curriculum

While each of the more than 200 BSCE curricula is designed for four years of study, it is possible to complete the four-year curriculum in three or three and a half years. On the other hand, students who need more courses before starting on the first calculus course in their freshman year need more than four years to complete. This is due to the fact that almost every BSCE program has prerequisites built into its curriculum. Prerequisites are necessary to ensure students taking a particular course have the required preparation because many of the courses do build on knowledge learned in other courses.

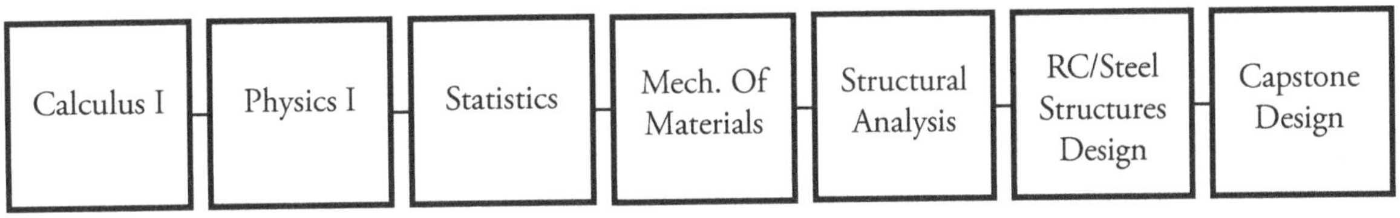


Figure 3.2 A key sequence of courses

A key sequence of courses that determines the minimum length of study is shown in Figure 3.2.

The above is for a curriculum requiring structural design in its capstone design course. Each preceding course is the prerequisite of the course following it. Even if the capstone design course in a particular curriculum does not include structural design, the requirements in another civil engineering technical area would produce another sequence of at least six courses. Thus, in order to graduate in four years or a shorter period, a student must not only pass all the key courses on the first try but also take these courses at the first possible opportunity. Your program has a curriculum flowchart for a 4-year plan, a 5-year plan, or a 6-year plan. You may have your own plan on how many courses you wish to take semester by semester. Do not be discouraged if you need more than four years to graduate. Many students need more than four years. In the assignments at the end, you are encouraged to work out your own flowchart with your advisor.

Before you graduate, there is one particular exam administered by the National Council of Examiners for Engineering and Surveying (NCEES) that you are encouraged to take and pass. The content of the exam is closely related to your curriculum. This is briefly introduced in the following section using contents published at the official NCEES site (www.ncees.org).

## 3.8 Fundamentals of Engineering Exam

The Fundamentals of Engineering (FE) Exam is the first step in the process leading to the professional engineer license. It is designed for students who are close to finishing an undergraduate engineering degree. The exam lasts 8 hours and is administered in April and October by the National Council of Examiners for Engineering and Surveying (NCEES). (This may change as NCEES is, as of 2013, discussing internally switching to an online exam format at different periods of a year.)

The FE exam contains 180 multiple-choice questions and is split into a morning session (120 questions) and an afternoon session (60 questions), each lasts four hours, with a one-hour lunch break in between.

The morning session is the same for everyone. For the afternoon session, you will be asked during registration to select the module (options are listed below) that best corresponds to your undergraduate degree:

Chemical Civil Electrical Environmental Industrial
Mechanical Other Disciplines

Let us take a look at the content of the **Morning Session**. It has 120 questions in 12 topic areas as listed below in detail with the approximate percentage of the test content following the topic area.

**I. Mathematics 15%**

A. Analytic geometry B. Integral calculus C. Matrix operations D. Roots of equations
E. Vector analysis F. Differential equations G. Differential calculus

**II. Engineering Probability and Statistics 7%**

A. Measures of central tendencies and dispersions (e.g., mean, mode, standard deviation)
B. Probability distributions (e.g., discrete, continuous, normal, binomial)
C. Conditional probabilities D. Estimation (e.g., point, confidence intervals) for a single mean
E. Regression and curve fitting F. Expected value (weighted average) in decision-making
G. Hypothesis testing

**III. Chemistry 9%**

A. Nomenclature B. Oxidation and reduction C. Periodic table D. States of matter
E. Acids and bases F. Equations G. Equilibrium H. Metals and nonmetals

**IV. Computers 7%**

A. Terminology (e.g., memory types, CPU, baud rates, Internet)
B. Spreadsheets (e.g., addresses, interpretation, "what if," copying formulas)
C. Structured programming (e.g., assignment statements, loops and branches, function calls)

**V. Ethics and Business Practices 7%**

A. Code of ethics B. Agreements and contracts C. Ethical vs. legal
D. Professional liability E. Public protection issues (e.g., licensing boards)

**VI. Engineering Economics 8%**

A. Discounted cash flow (e.g., equivalence, PW, equivalent annual FW, rate of return)
B. Cost (e.g., incremental, average, sunk, estimating)
C. Analyses (e.g., breakeven, benefit-cost) D. Uncertainty (e.g., expected value and risk)

**VII. Engineering Mechanics (Statics and Dynamics) 10%**

A. Statics: 1. Resultants of force systems, 2. Concurrent force systems, 3. Equilibrium of rigid bodies, 4. Frames and trusses, 5. Centroid of area, 6. Area moments of inertia, 7. Friction

B. Dynamics: 1. Linear motion (e.g., force, mass, acceleration, momentum), 2. Angular motion (e.g., torque, inertia, acceleration, momentum), 3. Mass moments of inertia, 4. Impulse and momentum applied to: a. particles, b. rigid bodies, 5. Work, energy, and power as applied to: a. particles, b. rigid bodies, 6. Friction

**VIII. Strength of Materials 7%**

A. Shear and moment diagrams B. Stress types (e.g., normal, shear, bending, torsion)
C. Stress strain caused by: 1. axial loads, 2. bending loads, 3. Torsion, 4. shear

D. Deformations (e.g., axial, bending, torsion) E. Combined stresses F. Columns
G. Indeterminate analysis H. Plastic versus elastic deformation

**IX. Material Properties 7%**
A. Properties: 1. Chemical, 2. Electrical, 3. Mechanical, 4. Physical
B. Corrosion mechanisms and control C. Materials: 1. engineered materials, 2. ferrous metals, 3. non-ferrous metals

**X. Fluid Mechanics 7%**
A. Flow measurement B. Fluid properties C. Fluid statics
D. Energy, impulse, and momentum equations E. Pipe and other internal flow

**XI. Electricity and Magnetism 9%**
A. Charge, energy, current, voltage, power
B. Work done in moving a charge in an electric field (relationship between voltage and work)
C. Force between charges D. Current and voltage laws (Kirchhoff, Ohm)
E. Equivalent circuits (series, parallel) F. Capacitance and inductance
G. Reactance and impedance, susceptance and admittance H. AC circuits
I. Basic complex algebra

**XII. Thermodynamics 7%**
A. Thermodynamic laws B. Energy, heat, and work C. Availability and reversibility
D. Cycles E. Ideal gases F. Mixture of gases
G. Phase changes H. Heat transfer I. Properties of enthalpy and entropy

Each of the 120 problems involves a single concept and counts for one point. You have on average two minutes for each problem.

The **Afternoon Session** for the **Civil Engineering Option** covers the following nine topic areas.

I. **Surveying 11%**
II. **Hydraulics and Hydrologic Systems 12%**
III. **Soil Mechanics and Foundations 15%**
IV. **Environmental Engineering 12%**
V. **Transportation 12%**
VI. **Structural Analysis 10%**
VII. **Structural Design 10%**
VIII. **Construction Management 10%**
IX. **Materials 8%**

We did not list the details for the Civil Engineering Option because many students prefer the **Other Discipline** option, which contains the following topics:

**I. Advanced Engineering Mathematics 10%**
A. Differential equations B. Partial differential calculus
C. Numerical solutions (e.g., differential equations, algebraic equations) D. Linear algebra
E. Vector analysis

**II. Engineering Probability and Statistics 9%**
A. Sample distributions and sizes B. Design of experiments C. Hypothesis testing
D. Goodness of fit (coefficient of correlation, chi square)
E. Estimation (e.g., point, confidence intervals) for two means

**III. Biology 5%**
A. Cellular biology (e.g., structure, growth, cell organization)
B. Toxicology (e.g., human, environmental)
C. Industrial hygiene (e.g., personnel protection equipment [PPE], carcinogens)
D. Bioprocessing (e.g., fermentation, waste treatment, digestion)

**IV. Engineering Economics 10%**
A. Cost estimating B. Project selection
C. Lease/buy/make D. Replacement analysis (e.g., optimal economic life)

**V. Application of Engineering Mechanics 13%**
A. Stability analysis of beams, trusses, and frames B. Deflection analysis
C. Failure theory (e.g., static and dynamic) D. Failure analysis (e.g., creep, fatigue, fracture, buckling)

**VI. Engineering of Materials 11%**
A. Material properties of: 1. Metals, 2. Plastics, 3. Composites, 4. Concrete

**VII. Fluids 15%**
A. Basic hydraulics (e.g., Manning equation, Bernoulli theorem, open-channel flow, pipe flow)
B. Laminar and turbulent flow C. Friction losses (e.g., pipes, valves, fittings)
D. Flow measurement E. Dimensionless numbers (e.g., Reynolds number)
F. Fluid transport systems (e.g., pipes, ducts, series/parallel operations)
G. Pumps, turbines, and compressors H. Lift/drag

**VIII. Electricity and Magnetism 12%**
A. Equivalent circuits (Norton, Thevenin) B. AC circuits (frequency domain)
C. Network analysis (Kirchhoff laws) D. RLC circuits
E. Sensors and instrumentation F. Electrical machines

**IX. Thermodynamics and Heat Transfer 15%**
A. Thermodynamic properties (e.g., entropy, enthalpy, heat capacity)
B. Thermodynamic processes (e.g., isothermal, adiabatic, reversible, irreversible)

C. Equations of state (ideal and real gases) D. Conduction, convection, and radiation heat transfer
E. Mass and energy balances F. Property and phase diagrams (e.g., T-s, h-P)
G. Tables of thermodynamic properties H. Cyclic processes and efficiency (e.g., refrigeration, power)
I. Phase equilibrium and phase change J. Thermodynamic equilibrium
K. Combustion and combustion products (e.g., CO, $CO_2$, NOX, ash, particulates)
L. Psychrometrics (e.g., humidity)

Comparing the topics covered by these two options, you will find that the Other Discipline option covers topics similar to the morning session topics but more in depth, while the Civil Engineering option covers seven civil engineering technical areas and civil engineering materials. Depending on how widely your curriculum covers these topics, you may choose one option against the other. In either case, the problems are more in depth and generally contain more than one concept and you have four minutes for each problem.

We list the topics herein not because you need to memorize them, but because you can compare the content with the course content of the corresponding course when you take it and pay due attention to the topics included in the FE exam. You are encouraged to update the content every year by visiting (www.ncees.org).

You are encouraged to take the exam as soon as you have taken all or most of the relevant courses in your program for the FE exam. Your program may offer a refresher non-credit course for you to prepare for the exam. Preparation is the key to passing the exam. In addition, the following data will encourage you the make every effort to pass on your first trial. According to NCEES, the pass rate for first-time takers of the Morning + Afternoon Civil Engineering Option was 67% and repeat takers 28% for the October 2012 FE exam. For the first-time takers of the Morning + Afternoon Other Discipline Option, the pass rate was 67% and repeat takers 30%. The percentage numbers do not add up to 100% because some exam takers did not provide the relevant information.

While the full scope of the FE exam is intimidating, you will be relieved to know that you are not expected to be able to answer all or even most of the questions. If every problem is answered correctly, one would receive 240 points (120×1 + 60×2). According to the statistics of NCEES, one would pass the exam if one scores in the neighborhood of 110 points. Furthermore, there is no penalty for a wrong answer. You know what that means.

How passing the FE exam will eventually lead to a professional engineer license will be described in Chapter 7.

## Additional Reading Recommendations

1. Read your university's online or print catalog to find out the mission statements of your university and the program objectives of your BSCE program. In the context of ABET evaluation, the program objectives are to be achieved by graduates of the program in several years.
2. *Studying Engineering: A Road Map to a Rewarding Career*, Raymond B. Landis, 4th edition, Discovery Press (www.discovery-press.com), 2013. Chapter 2 of this book contains a description of all engineering disciplines, including civil engineering.

## Assignments

1. Use the curriculum requirements of your BSCE program to produce your own Degree-Progress-Plan. Take into account (a) prerequisite compliance, (b) time available to you for studying as determined from your assignment in Chapter 2, and (c) the advice from your faculty advisor. Use course number only. For general education and non-specifically required courses fill in the category and credit hours

**Table 3.1 Planned courses to take toward graduation**

| | | | | | | |
|---|---|---|---|---|---|---|
| Required Courses passed, transferred, and placed out as of the beginning of the current semester | | | | | | |
| | | | | | | |
| | | | | | | |
| | | | | | | |
| Remaining required courses including the courses you are taking | | | | | | |
| | | | | | | |
| | | | | | | |
| | | | | | | |
| | | | | | | |
| | | | | | | |
| | | | | | | |
| | | | | | | |
| | | | | | | |
| | | | | | | |
| Semester | Course 1 | Course 2 | Course 3 | Course 4 | Course 5 | Total Credit-Hours |
| Fall 20____ | | | | | | |
| Spring 20___ | | | | | | |
| Summer ___ | | | | | | |
| Fall 20____ | | | | | | |
| Spring 20___ | | | | | | |
| Summer ___ | | | | | | |
| Fall 20____ | | | | | | |
| Spring 20___ | | | | | | |
| Summer ___ | | | | | | |
| Fall 20____ | | | | | | |
| Spring 20___ | | | | | | |
| Summer ___ | | | | | | |
| Fall 20____ | | | | | | |
| Spring 20___ | | | | | | |
| Summer ___ | | | | | | |
| Fall 20____ | | | | | | |
| Spring 20___ | | | | | | |
| Summer ___ | | | | | | |

Table 3.2 Analysis of the Five Elements in My Civil Engineering Curriculum

| Five Elements | | | | | | | | | Credit-Hours |
|---|---|---|---|---|---|---|---|---|---|
| General Education | | | | | | | | | |
| | | | | | | | | | |
| Math & Sciences | | | | | | | | | |
| | | | | | | | | | |
| ES & CE Fundam.* | | | | | | | | | |
| | | | | | | | | | |
| CE Tech Areas | | | | | | | | | |
| | | | | | | | | | |
| Capstone | | | | | | | | | |
| Total | | | | | | | | | |

* Engineering Sciences and Civil Engineering Fundamentals

associated with each category. For the semester-by-semester plan, start with the current semester and the courses you are taking.

2. Create your Degree Portfolio, a file containing (a) your curriculum flowchart, (b) your course-outcomes matrix (ask your advisor to provide you a copy of the matrix correlating the ABET EC2000 outcomes to each of the required courses of your program), (c) your Degree-Progress Plan as indicated in the above assignment, and (d) the FE exam Morning Session topics and Afternoon Session topics. Update the portfolio every semester.
3. Analyze your curriculum by filling in the course numbers in the following table. For general education and non-specifically required courses, fill in the category and credit-hours associated with each category.
4. How many civil engineering technical areas are required in your curriculum? List the courses associated with each area.
5. In the ABET EC2000, how many attributes as student learning outcomes are required for graduating engineering students to have acquired: (a) 8, (b) 9, (c) 10, (d) 11, or (e) 12?
6. How many student learning outcomes beyond the number of ABET EC2000 learning outcomes are in your civil engineering program? (a) 0, (b) 1, (c) 2, (d) 3, or (e) more than 3.
7. Does your BSCE program require additional student learning outcomes beyond the ones specified in ABET EC2000? If it does, specify.
8. ABET accredits (a) engineering, technology, and engineering science programs, (b) engineering and technology schools and colleges, (c) universities with engineering and/or technology programs, (d) online engineering and/or technology courses, or (e) none of the above.
9. Who actually designs the BSCE program you are in: (a) ABET, (b) your university's academic vice president, (c) your college dean, (d) your department chair, or (e) your program faculty?
10. How many mathematics courses are required in your BSCE curriculum and how many credit-hours are associated with each course?
11. How many basic sciences courses are required in your BSCE program and how many credit-hours are associated with each course?

12. The FE exam includes topics typically covered in (a) freshman year, (b) sophomore year, (c) junior year, (d) senior year, or (e) all three years from freshman to junior.
13. Included in the engineering science courses in your curriculum are (a) statistics, (b) dynamics, (c) thermodynamics, (d) fluid dynamics, or (e) all of the above.
14. The very first course in the key sequence of a BSCE curriculum is (a) calculus I, (b) physics I, (c) statics, (d) dynamics, or (e) chemistry I.
15. A unique feature in the BSCE curriculum requirement of ABET, in comparison to other engineering programs, is the requirement of one additional (a) mathematics course, (b) science course, (c) engineering science course, (d) design course, or (e) all of the above.
16. Why must your BSCE program be accredited? Because accreditation ensures (a) academic quality, (b) accountability, (c) facilitate course credit transfer, (d) scrutiny for needed change, or (e) all of the above.
17. In recent years, the passing rate for first time exam takers of the FE exam has been close to (a) 40%, (b) 50%, (c) 2/3, (d) 3/4, or (e) 80%.
18. In recent years, the passing rate for repeated exam takers of the FE exam has been close to (a) 30%, (b) 40%, (c) 50%, (d) 60%, or (e) 70%.
19. Your BSCE program is accredited by ABET. How often does ABET review your program? Every (a) 2 years, (b) 3 years, (c) 4 years, (d) 5 years, or (e) 6 years.

# Chapter 4
# Co-Curricular Learning

## 4.1 Overview

Learning by taking courses is the core of college education for many years, but students on any campus also have opportunities to participate in a variety of campus-based activities other than taking courses. These activities used to be characterized as extracurricular. In recent years educators increasingly view these extracurricular activities as complementary to the curricular learning and called such activities as co-curricular learning instead of extra-curricular to emphasize its usefulness to students. These co-curricular activities are organized by campus-sanctioned student organizations. For civil engineering students there are three types of student organizations of interest: civil engineering–related organizations, engineering-related organizations, and other organizations. A civil engineering student may choose to participate in activities of an organization not related to civil engineering or engineering because of personal preference or hobby. Whether or not an organization has anything to do with civil engineering, the common benefits to a participant are:

1. It provides **a sense of belonging**. Members of a student organization form a small community within the larger community of a campus. For active participants it becomes a home away from home. Educators believe a sense of belonging is a factor in encouraging learning and overcoming academic challenges.
2. It provides an environment for **improving interpersonal skills**. Rarely does a civil engineer work alone. Whether one works with his/her colleagues in a firm or a government agency, dealing with people is a daily occurrence. Interaction within a student organization provides the opportunity for an active participant to learn from mistakes and successes in dealing with people.
3. It provides a training ground for **teamwork**. Most student organizations have group activities. Active participants learn to be **good leaders and good followers**. Followership is as important as leadership in making a team function properly and in personal development.
4. It provides an opportunity to **know people from other majors**. To a civil engineering student, this is true for non-civil engineering organizations. Knowing people from other majors broadens one's perspective on issues concerning the larger community of a society.

5. It provides an opportunity in **networking with people from the outside world**. This is true for the participants of the student chapter of a national organization. Typically such a student chapter has one or more advisors from the national organization and has regional and national activities sanctioned by the national organization. These outside advisors may provide career mentorship as well as academic advisement.
6. It provides scholarships to qualified members.

For an active member of a student organization, time spent on the activities of the organization is time not spent directly on studying for courses. The obvious question is if such activities hurt the academic performance of a student. That depends on how a student manages time. Students active in student organizations usually manage their time wisely. Of course if excess amount of time is spent on a student organization, then it will hurt one's academic performance. One useful rule of thumb is: time spent on activities of a single student organization is equivalent to that of taking one 3-credit course. One should manage time accordingly.

The following student organizations are most relevant and most likely to be joined by civil engineering students. Following the description of these relevant organizations, other co-curricular learning opportunities are introduced.

## 4.2 ASCE

**ASCE Overview**. Founded in 1852, the American Society of Civil Engineers is the oldest national engineering society. Presently its local organizations are organized in nine US regions and an international region. Each region is further organized in sections, and within each section, branches.

ASCE serves its members and the public interest. An example of its commitment to public interest is its annual report by its Committee on Critical Infrastructure (CCI) focusing on areas related to critical infrastructure resilience. Its Report Card for America's Infrastructure calls the public's and political leaders' attention to the urgent need in infrastructure renewal.

The Society has eight Institutes created to serve professionals working within specialized fields of civil engineering:

Architectural Engineering Institute (AEI)
Coasts, Oceans, Ports and Rivers Institute (COPRI)
Construction Institute (CI)
Engineering Mechanics Institute (EMI)
Environmental and Water Resources Institute (EWRI)
Geo-Institute (G-I)
Transportation and Development Institute (T&DI)
Structural Engineering Institute (SEI)

ASCE's Technical Activities Committee (TAC) has ten Divisions and Councils:

Aerospace Engineering
Cold Regions Engineering

Computing in Civil Engineering
Energy
Forensics Engineering
Geomatics
Lifeline Earthquake Engineering
Disaster Risk Management
Pipelines and Sustainability
Technical Activities Committee

ASCE is the largest publisher of civil engineering technical and professional information. Its thirty-three journals are the premier source of technical information for the civil engineering profession.

In addition to these journals, ASCE also publishes books and the ***ASCE News*** and ***Civil Engineering*** magazine, which is an important resource for nation-wide job postings in industry and academe.

ASCE annually honors achievements of its members for important contributions to the technical and professional areas with numerous technical and professional awards, and presents the awards in public meetings or conferences. These awards are considered the most prestigious in the profession.

## 4.3 ASCE Student Chapter Activities

Student chapters of ASCE are generally associated with BSCE programs, but some community colleges also host ASCE student chapters if they offer an associate degree in civil engineering. To maintain good standing, a student chapter of ASCE must submit an annual report on its activities to show that its members have indeed participated in a variety of technical activities and community service projects. Specific benefits to student chapter members include free online access to the *Civil Engineering* magazine and *ASCE News*, access to career resources such as information on internship and job opportunities, professional guidance through the chapter's advisors and ASCE's mentoring program, and networking opportunities through annual conferences and competitions.

For student chapter members, ASCE offers a free national student membership, but students do pay dues to their own local student chapter. Each student chapter has three advisors: one Faculty Advisor and two Practitioner Advisors formerly referred to as "Section Contact Members." The Faculty Advisor is a member of the civil engineering faculty. The Practitioner Advisors are members of an ASCE Section or Branch and provide contact between the students and the civil engineering professionals in the area. These individuals provide guidance to the student leaders, help maintain a balance in group activities, and foster the development of professionalism and leadership in the members of the ASCE group. The Faculty Advisor is selected by the Civil Engineering Department Head. The Practitioner Advisors are selected by the ASCE Section or Branch.

To provide guidance to the student chapters, ASCE has a Committee on Student Activities (**CSA**) with ten civil engineering educators and practitioners who help guide the overall national ASCE student program.

## 4.4 ASCE Student Awards and Competitions

Awards and competitions available to individuals or groups are briefly described here. The concrete and steel bridge competitions are described in more detail because they are the most popular annual activities participated in by the most student chapters.

**Student Organization Awards**: Based on activities recorded in the annual report of the student chapters, the CSA determines the recipients of awards honoring student organizations, faculty advisors, and practitioner advisors:

1. **Robert Ridgway Student Chapter Award**: This award was endowed by Isabel L. Ridgway in honor of her husband, Robert Ridgway, 1925 national president, ASCE. It was officially instituted in May 1965. The award is made annually to the single most outstanding Student Chapter of the ASCE. Several chapters are also honored as the Ridgway Award Finalists.
2. **Regional Governors Award**: The CSA selects the best student chapter from each of the nine domestic regions and the international region to receive the Regional Governors award.
3. **Richard J. Scranton Outstanding Community Service Award**: This award honors the student chapter with the most outstanding community service activities.
4. **Other Student Chapter Awards**: **Certificates of Commendation**, **Letters of Honorable Mention**, **Most Improved Award**, **Letters of Significant Improvement**: There are non-student awards: One faculty from each of the ten ASCE regions is selected to receive the **Outstanding Faculty Advisor Awards**. Numerous outstanding faculty advisors are selected for the **Faculty Advisor Certificates of Commendation**. One or more practitioner advisors are selected for the **Outstanding Practitioner Advisor Awards**. One or more practitioner advisors are selected for the **Practitioner Advisor Certificates of Commendation**.

**Daniel W. Mead Student Contest**: This is a single essay competition on a topic selected annually by the CSA. The contest is established in 1939 in honor of ASCE's 67th president, Daniel W. Mead, and to further young civil engineers' professional development. Only one essay is allowed to enter the competition from each student chapter and is to be submitted through the faculty advisor. The faculty advisor becomes the de facto arbitrator on selecting the best essay out of the chapter for the national competition. Up to five winners receive cash awards, ranging from $1,000 to $200.

**Student Leadership Award**: This award is granted to someone with demonstrated leadership in terms of serving as chapter officers, leading chapter activities, interacting with university administration, and interacting with ASCE branches/sections. Candidates for this award must be nominated by the faculty or practitioner advisor, department head, or branch/section officers.

**ASCE National Concrete Canoe Competition**: The use of concrete in floating vessels can be traced back to Joseph-Louis Lambot who built concrete boats for use on his estate in France in 1848. ASCE has been sponsoring this national competition since 1988 when the then-Masters Builder offered to become the sole corporate sponsor. ASCE student chapter activities in concrete canoe, however, had been in existence for more than 20 years by 1988. The idea of a national competition was initiated in 1985 by Dr. R. John Craig, a professor of the New Jersey Institute of Technology. By 1987 the rules were formulated and approved but Dr. Craig was diagnosed with a rare form of brain tumor. He passed away only two months before his vision of a national concrete canoe competition became reality in 1988.

The objectives of the national competition are described by ASCE as follows:

- To provide civil engineering students an opportunity to gain hands-on, practical experience and leadership skills by working with concrete mix designs and project management.
- To increase awareness of the value and benefits of ASCE membership among civil engineering students and faculty in order to foster lifelong membership and participation in the Society.
- To build awareness of the versatility and durability of concrete as a construction material among civil engineering students, educators and practitioners, as well as the general public.
- To build awareness of concrete technology and application among civil engineering students, educators and practitioners, as well as the general concrete industry.
- To increase awareness among industry leaders, opinion makers, and the general public of civil engineering as a dynamic and innovative profession essential to society.
- To generate and increase awareness of all sponsoring companies' commitment to civil engineering education among civil engineering students, educators and practitioners, as well as the general public.

The competition generally includes the design and construction of a concrete canoe, a paper presentation and an oral presentation, and five types of races. Over the years ASCE has developed clearly defined rules and regulations governing the competition. Entry to the national competition, which is hosted by a different school every year, is limited to the top two schools (winners) of the eighteen student conference competitions, plus the host team. The same canoe must be used in both the conference and the national competitions. Some of the rules and regulations and the components of the competition are highlighted below.

Figure 4.1 The co-ed concrete canoe sprint race at the 2008 National Concrete Canoe Competition in Montreal, Quebec.

Source: http://en.wikipedia.org/wiki/File:Coed_Sprint.jpg. Copyright in the Public Domain.

a. **Materials**: By definition concrete canoes are made of concrete with reinforcement. For the competition, there are strict specifications on the material used. The concrete used must be designed by the team. Prepackaged or premixed concrete, mortar, or grout is not permitted. Each of the four components of the concrete material: cement (including other cementitious materials such as fly ash or silica fume), aggregate, fiber, and admixture are subject to a variety of established requirements.
b. **Size and Shape**: The length of the beam width and the depth are specified every year. While the specified dimensions cannot be changed, the designer has the discretion of determining the hull thickness and the dimension of any thwarts and ribs.
c. **Design Paper**: Each team must submit a design paper detailing the design and construction of the concrete canoe, including the project management and the innovations and sustainable aspects of the design. The paper shall include an organization chart with names of team members and their tasks and contributions, a project schedule, a design drawing, and bill of materials form, among other required contents.
d. **Engineer's Notebook**: The Engineer's Notebook is a technical document providing supportive information on the design and construction of the concrete canoe. It includes photographs of the construction of the canoe at various stages, hull thickness and reinforcement-related calculations, and technical datasheet on the products used in the concrete canoe.
e. **Oral Presentation**: Each team is to make a five-minute live technical presentation on the various aspects of the project, followed by seven minutes of questions and answers. Video is permitted but no pre-recording of the speaking parts is allowed.
f. **Final Product: Canoe and Cutaway Section Display**: In addition to the finished concrete canoe, a cutaway section of at least three feet is to be displayed alongside the canoe. The cutaway section shall demonstrate the concrete casting, finishing, and the reinforcement techniques used.
g. **Canoe Floating Test**: To assure safety and durability, the concrete canoe is required to survive a floating test. The canoe is placed in water and filled up inside with water. To pass the floating test, the point within 2 feet and 6 inches of the two extreme ends of the canoe must break the water surface simultaneously within five minutes of being completely filled with water.
h. **Races**: A total of five races are held ranging in distance from 200 m to 600 m: women's slalom/endurance (3 women), men's slalom/endurance (3 men), women's sprint (2 women), men's sprint (2 men), and co-ed sprint (2 women, 2 men). All paddlers shall be competent swimmers and shall wear U.S. Coast Guard-approved life jackets at all times while they are in the canoe. A powered rescue boat is required to be present during all races.

Seventy-five percent of the total team score is based on engineering design and construction principles, the written report, and oratory skills. The remainder is based on the performance of the canoe and the paddlers in five different race events. Based on the overall scores, scholarship awards and trophies are given to the three teams with the highest scores: $5,000, $2,500, and $1,500, respectively. Special plaques are given to the fourth- and fifth-place teams, best design paper, best oral presentation, best final product, spirit of competition, and each of the winners of the five races. The winner of the co-ed sprint race receives a special plaque honoring the founder of the national race, Dr. R. John Craig.

One of the challenges of the concrete canoe competition is the transportation of the canoe itself. Many times, the canoe is damaged during transportation and handling to the point that further participation becomes untenable. Despite the potential problem of failure, the competition remains a popular one because

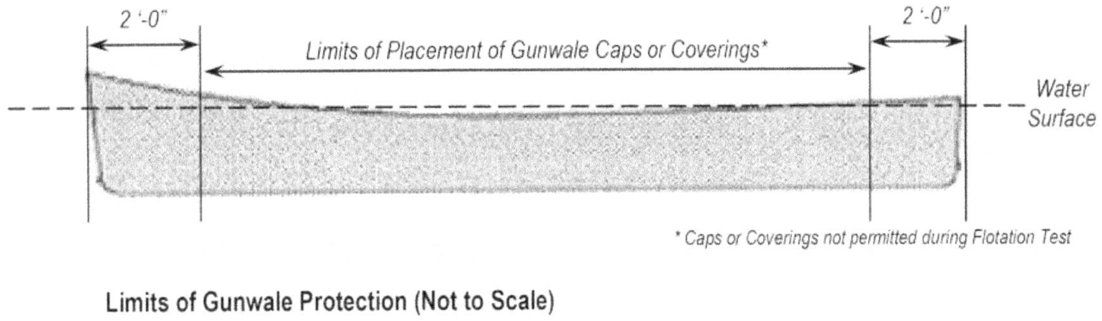


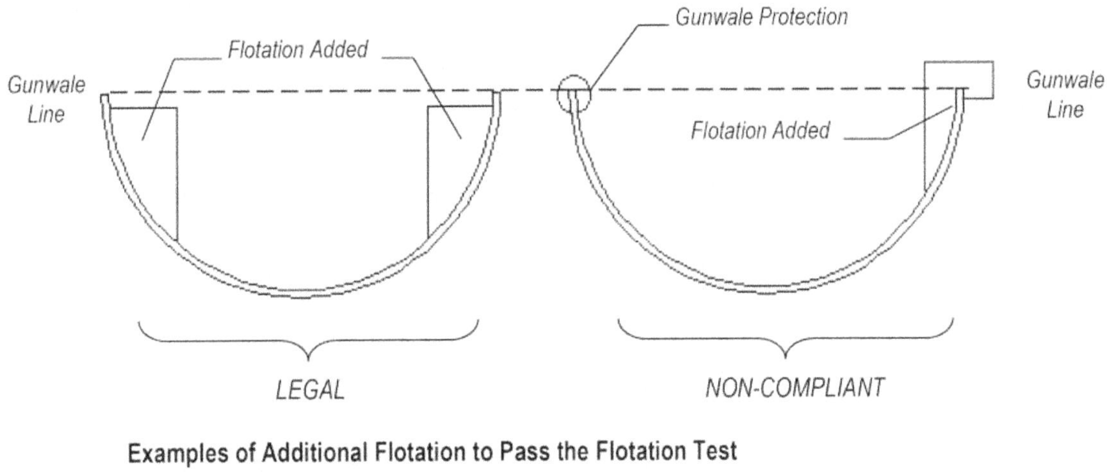


Figure 4.2 Two Figures from the 2013 Official Rule book.

Figure 2.2, 2013 Rules and Regulations, pp. 8. Copyright © 2013 by American Society of Civil Engineers (ASCE). Reprinted with permission.
Figure 2.3, 2013 Rules and Regulations, pp. 8. Copyright © 2013 by American Society of Civil Engineers (ASCE). Reprinted with permission.

it entails a wide range of activities challenging participants' minds and bodies. The above description is usually changed from year to year in its details but not in its general outline. You are encouraged to find the most up-to-date information at the ASCE site: (http://www.asce.org/concretecanoe/).

**ASCE/AISC Student Steel Bridge Competition**: This national competition is sponsored by ASCE and American Institute of Steel Construction (AISC) and co-sponsored by several steel industry organizations or companies. AISC is a not-for-profit technical institute and trade association established in 1921. Its mission as described by AISC is "to serve the structural steel design community and construction industry in the United States. AISC's mission is to make structural steel the material of choice by being the leader in structural-steel-related technical and market-building activities, including: specification and code development, research, education, technical assistance, quality certification, standardization, and market development."

Teams enter the competition first at the regional level through the ASCE Student Conferences. Up to three teams from the winners of each of the regional competitions are qualified to enter the national competition. The rule of the competition is changed every year to enhance the competitiveness. Some of the main features of the competition are described below.

a. Material and Size: Obviously only steel is used. The steel bridge's dimension, the sizes of each member of the bridge, and the nuts and bolts connected by steel fasteners specified/limited.
b. Display and Timed Construction: The bridge is first displayed. Design of the bridge is to be described on a poster displayed with the bridge. The appearance of the bridge is judged by its balance, proportion, elegance, and finish. After the display, the bridge is disassembled. It is assembled again on site at a designated location provided by the host school during the timed construction competition for construction speed. Time of assembling is the main factor of this competition. A penalty is assessed if the assembling time exceeds 30 minutes. In addition, the number of "builders" participating in the construction is also a factor in assessing the overall construction economy.
c. Weight and Stiffness: Obviously the lighter the better, but it is not the weight alone that is judged. Various penalties are assessed in terms of added weight. During the load tests, three points of the bridge are measured for their vertical deflection. The aggregated deflection amount reflects the stiffness of the bridge.
d. Loading and Unloading: A lateral load test is conducted and the lateral sway amount is measured. If excessive sway is observed, the bridge is deemed unsafe and no further tests are to be conducted. If a bridge passes the lateral load test, the vertical load test is conducted. The vertical loading test actually has a preload phase and a loading phase, and then the loads are taken off. If a bridge collapses during the unloading phase, the team will not receive any awards.
e. Awards: Seven categories of performance are used for competition: display, construction speed, construction economy, lightness, stiffness, structural efficiency, and overall performance. The top three teams in each category are recognized with a plaque. For the national competition each team is provided a cash travel subsidy.

During the construction and loading of the bridge, safety is of major concern. The construction is not as easy as one might imagine; there are rules governing the behavior of all the builders. For example, no builder can use any other builder for support. No builder is allowed to cross the line of "floodway" of the "river." Sometimes the builders twist their body to reach the target bridge. Falling and twisted angles may result from overextended postures. Despite these potential hazards, the excitement of putting together a bridge designed by students themselves always attracts many participants. For updates, check the official website:

http://www.aisc.org/WorkArea/showcontent.aspx?id=21576.

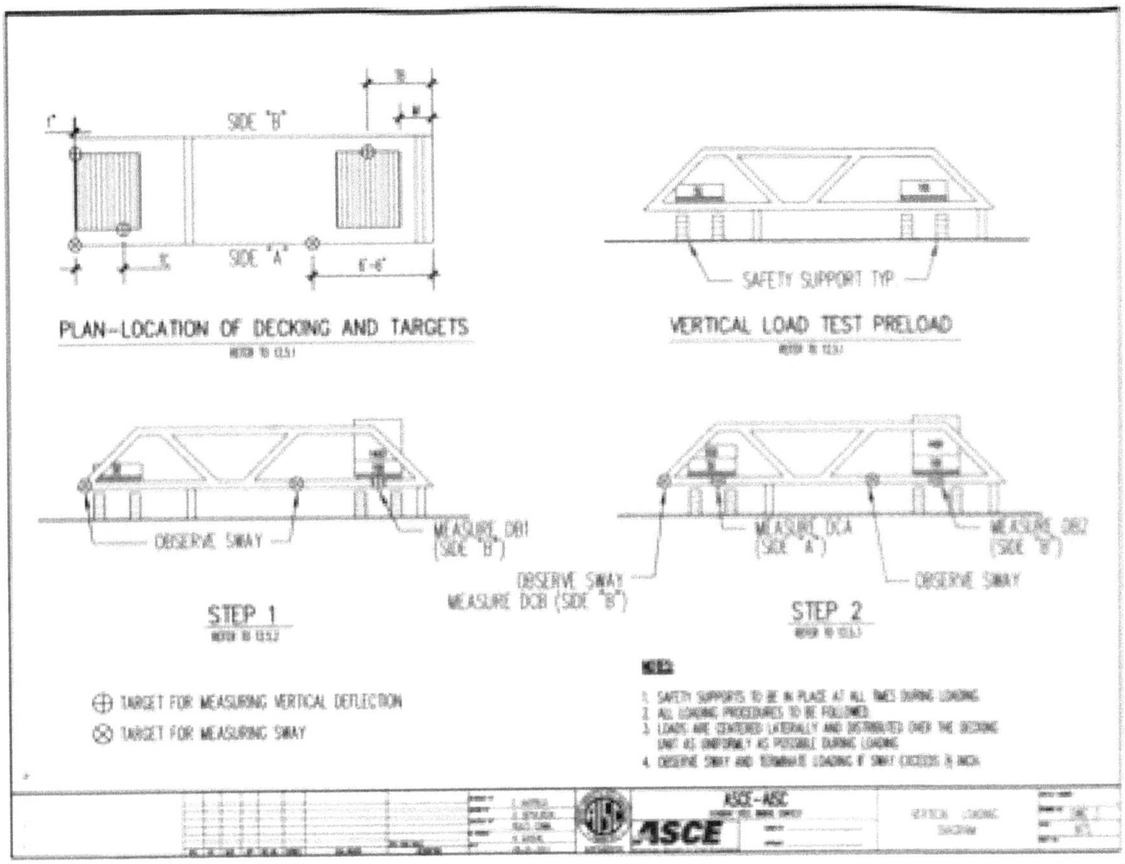


Student Steel Bridge Competition: 2013 Rules, pp. 40. Copyright © 2013 by American Society of Civil Engineers (ASCE). Reprinted with permission.

## 4.5 Student Chapter of Chi Epsilon: Civil Engineering National Honor Society

Many BSCE programs have a student chapter of Chi Epsilon, the national honor society for civil engineering students, faculty, and alumni. The society, as described by Chi Epsilon, is "Dedicated to the purpose of maintaining and promoting the status of civil engineering as an ideal profession, Chi Epsilon was organized to recognize the characteristics of the individual civil engineer deemed to be fundamental to the successful pursuit of an engineering career, and to aid in the development of those characteristics in the civil engineering student." The society's history can be traced back to 1922 when two fraternities at University of Illinois were independently established. A national society was established in 1923. By 1958 the national society had 49 chapters. It grows at a rate of 20 chapters every decade. By 2008, there were 233 chapters.

As an honor society it initiates student members by their scholastic achievements. In other words, not every civil engineering student can apply to become a member. For civil engineering undergraduates, only juniors and seniors in the top one-third of their class (typically ranked according to GPA) are eligible.

To be a member of Chi Epsilon is an honor for a civil engineering student. They are also eligible to compete for numerous scholarships from $1,500 to $3,000. In case your institute does not have a student chapter of Chi Epsilon, you can initiate one. Check with your department chair and official site: (http://xe.uta.edu/xewebgeneral2/).

## 4.6 Student Chapter of Tau Beta Pi, Engineering Honor Society

While Chi Epsilon is a civil engineering honor society for civil engineers only, Tau Beta Pi is a national honor society for all engineering disciplines. As described by Tau Beta Pi, its mission is "to mark in a fitting manner those who have conferred honor upon their Alma Mater by distinguished scholarship and exemplary character as students in engineering, or by their attainments as alumni in the field of engineering, and to foster a spirit of liberal culture in engineering colleges. Its vision: Tau Beta Pi will be universally recognized as the premier honor society. Its creed: Integrity and excellence in engineering." Its history traces back to 1885 when it was founded at Lehigh University. By 2009 it had 241 chapters.

Eligibility requirements for undergraduate students include scholastic achievements measured by ranked in the top one-eighth of their major for juniors and top one-fifth for seniors. Members are eligible to compete for a variety of scholarships.

## 4.7 Other Student Chapters

There are other student chapters of national societies on many campuses that civil engineering students are likely to join. Only a few engineering societies are briefly described below, in the order of the year they were founded.

**AWWA, American Water Works Association**. Founded in 1881 by 22 men, the purpose of the association has been "for the exchange of information pertaining to the management of water-works, for the mutual advancement of consumers and water companies, and for the purpose of securing economy and uniformity in the operations of water-works." AWWA has more than 50,000 members working to protect public health and water resources for future generations and 41 student chapters on campuses in U.S.A., Canada, and Mexico (www.awwa.org). While many BSCE programs offer introductory undergraduate courses related to drinking water, the majority of drinking water courses are usually offered at the graduate level. As a result student chapters of AWWA are populated mostly by graduate students. The scholarship program also concentrates on M.S.- and Ph.D.-level students, and so is its student competition program.

**ACI, American Concrete Institute**. Founded in 1904, ACI has become the premium organization advancing concrete technology in the world. Its detailed specifications on concrete design and application are adopted into national and local government codes. ACI offers the following general description: "With 99 chapters, 37 student chapters, and nearly 20,000 members spanning 108 countries, the American Concrete Institute has always retained the same basic mission—to develop, share, and disseminate the knowledge and information needed to utilize concrete to its fullest potential."

ACI student members are active participants in ACI's activities and student competitions even without a student chapter. Self-organized student teams can participate in the following competitions:

a. **ACI FRC Bowling Ball Competition**: The object is to design and construct a fiber-reinforced concrete bowling ball to achieve optimal performance under specified failure criteria and to develop a fabrication process that produces a radial uniform density while maximizing volume.
b. **ACI Concrete Cylinder Competition**: The objective is to produce concrete cylinders with an average compressive strength of 7,000 psi (48.3 MPa) and a saturated surface-dry density of 150 lb/ft$^3$ (2.39 kg/l) with the highest cementitious efficiency and the lowest cost, and to write a report explaining the design and production process.
c. **ACI Concrete Projects Competition**: Virtually any project that focuses on concrete design, materials, and/or construction is eligible. These projects can include computer programs, term papers, student activities, senior design projects, or special projects.
d. **ACI FRP Composites Competition**: Students design, construct, and test a concrete structure reinforced with fiber-reinforced polymer (FRP) bars to achieve the optimal load-to-weight ratio, predict the ultimate load, and predict the load that will result in a piston deflection of 2.5 mm (0.1 inch).
e. **ACI Concrete Cube Competition**: Students produce a concrete cube that achieves, as closely as possible, a target design strength of 50 MPa and a target mass of 270 grams per cube.
f. **ACI Egg Protection Device Competition**: Students design and build the highest-impact load-resistant plain or reinforced concrete Egg Protection Device.
g. **ACI Concrete Construction Competition**: The Concrete Construction Competition is for undergraduate students with interests in construction technology, construction management, and concrete industry management. Student teams (comprising up to 5 students each) are given one week to provide a response to a realistic, open-ended question on the subject of concrete construction.

Through its non-profit subsidiary, ACI Foundation, ACI offers fellowships and scholarships every year. The scholarships are for graduate students only. The fellowships are for undergraduate and graduate students with educational stipends to cover tuition, residence, books, and materials valued at $7,000 to $10,000. Candidates must be nominated by a faculty who must be a member of ACI. A 500-word essay is required along with transcripts and two references submitted online (www.concrete.org).

**SWE, Society of Women Engineers**. Founded in 1950, SWE is "a not-for-profit educational and service organization that empowers women to succeed and advance in the field of engineering, and to be recognized for their life-changing contributions as engineers and leaders." SWE has about 400 professional and collegiate sections throughout ten regions in the United States (www.swe.org).

SWE has a very active scholarship program. In 2012, SWE disbursed 198 new and renewed scholarships valued at $577,000. Recipients include freshmen through graduate students. Individual scholarships range from $1,000 to $10,000. Many local professional sections also award scholarships.

The student sections are encouraged to participate in the SWE's collegiate competition program. The available competitions include

a. Outstanding Collegiate Section Award: It recognizes the best sections for their overall performances. Its gold level award typically recognizes five student sections.
b. Outstanding New Collegiate Section Award: It recognizes the best of the new sections.

c. Collegiate Technical Poster Competition: it recognizes the best ability to deliver visual presentations.
d. Team Tech Competition: Sponsored by the Boeing Company, this award emphasizes the importance of teamwork and interface with industry in the engineering educational process.
e. SWE Bowl: Sponsored by the ExxonMobil Company, the Subject Matter Expert (SME) Bowl is a competition that challenges collegiate students to show their stuff in the areas of science, math, physics, engineering, and technology. It also challenges these students in their knowledge of SWE history and policies.

**NSBE, National Society of Black Engineers**. Started as a student organization at Purdue University in 1971, NSBE became a national non-profit organization in 1976. It has more than 250 student chapters. NSBE's mission (www.nsbe.org) is "to increase the number of culturally responsible Black Engineers who excel academically, succeed professionally and positively impact the community."

NSBE has a wide range of programs: collegiate programs, pre-college initiative programs, graduate student programs, technical professional programs, community outreach and service programs, and scholarship programs.

For undergraduate students, NSBE offers seventeen NSBE and corporation-sponsored scholarships ranging in value to individual recipients from $500 to $7,500.

**SHPE, Society of Hispanic Professional Engineers**. Founded in 1974 by several professional engineers in Los Angeles, SHPE has quickly grown into a very active national organization. More than 250 local and student chapters on engineering campuses are organized in seven national regions.

As described by SHPE, the society's mission: "SHPE changes lives by empowering the Hispanic community to realize their fullest potential and impacts the world through STEM awareness, access, support and development." Vision: "SHPE's vision is a world where Hispanics are highly valued and influential as the leading innovators, scientists, mathematicians and engineers." Student members are eligible to compete for internships, scholarships, and awards.

The SHPE has an extensive scholarship program offered through the **Advancing Hispanic Excellence in Technology, Engineering, Math and Science (AHETEMS) Foundation,** which is an independent non-profit organization working exclusively to develop educational enrichment and academic outreach initiatives, for Latinos/as that extends throughout the pre-college to Ph.D. pathway. Several types of scholarship are shown below:

a. AHETEMS General Scholarships: These merit-based and need-based general scholarships are awarded, in the amount of **$1,000–$3,000**, to qualified high school graduating seniors, undergraduate students, and graduate students who demonstrate both significant motivation and aptitude for a career in science, technology, engineering, or mathematics.
b. AHETEMS/ExxonMobil Scholarship: These merit-based scholarships are awarded in the amount of **$2,500 to undergraduate students** who demonstrate both significant motivation and aptitude for a career in science, technology, engineering, or mathematics.
c. AHETEMS/Kellogg Corporation Scholarship, Northrop Grumman Scholarships, U.S. Steel Corp. Scholarships, Verizon Scholarships: Each of the four scholarships is merit-based and awarded in the total amount of **$5,000 to undergraduate students** who demonstrate both significant motivation and aptitude for a career in engineering.

SHPE sponsors student competitions for undergraduates and makes its awards to winners during its annual conference:

a. Student Technical Paper Competition: Participants submit abstracts and full papers (if abstracts are accepted), and make presentations during the conference. Winners are selected based on papers' originality, social impact, and completeness.
b. Student Technical Poster Competition: Participants submit abstracts and full posters (if abstracts are accepted), and avail themselves during the poster session of the conference. Winners are selected based on posters' originality, social impact, and completeness.
c. Academic Olympiad and Design competition are sometimes sponsored.

## 4.8 Other Co-Curricular Options

### Undergraduate Research Experience (URE)

**Benefits**. Participating in a research program as an undergraduate has many benefits. You will observe up close how research is conducted. You have the opportunity to work closely with a faculty. Most likely you will interact with other students (graduate research assistants and/or other undergraduates) on a team. Your experience gives you advantage if and when you apply for graduate study and research assistantships. You will be paid for your work.

**Duration**. The duration of your participation often depends on the funding available and your availability. It could be a summer program, a semester program, or a year-long program. You could work full time during the summer and part time during the school year.

**Place of Research**. The place of conducting research usually is on your own campus, but many universities offer undergraduate research programs on an open-competition basis. It usually is a summer program. There will be an open application announcement sent to your university or your department by the host institute inviting applications. You apply according to the requirements announced. If accepted, you will then spend a summer at another institute, giving you the additional benefit of experiencing a new campus and a new location.

**Funding**. Usually you will work for a faculty who has a funded research project with a budget that pays you for your work. A major funding agency, the National Science Foundation (NSF) has a long-running program for undergraduate research experience. Any faculty director of an NSF-funded project can apply for supplemental funding for URE and has a high probability of getting approved. NSF also funds URE at an institute that supports a group of undergraduate students from other campuses as described above.

**Where to Apply**. For URE on your own campus, you need to watch out for announcements and seek help from your faculty advisor. You can also ask each faculty who has a funded research program for the opportunity. For opportunities at other campuses, check with your faculty advisor or department chair in spring or late winter. These opportunities are usually for summer only as described above.

**Keep a Record**. It will benefit you to keep a daily work log. It includes a description of your assignment and how the assignment is fulfilled. You may add any thoughts on your work of that day. It is like a diary but devoted to the URE only. The daily log shall follow an overview describing your responsibilities and the project you are in. At the conclusion of the URE period, you can add a summary on your overall experience. Such a record will come in handy when you apply for graduate study and assistantship and when you apply for a job.

## Design Clinics

**Benefits.** Design clinics are funded by industrial sponsors for a faculty-student team to solve a particular problem for the sponsor. The participating students usually are paid to work on the real-world problem. Students have all the benefits of an undergraduate research experience project but work on a problem more on the development side than the research side.

**Duration.** It depends on the nature of the problem and funds available; it ranges from a summer or a semester to more than a year. Students typically work on a part-time basis during semesters and full time in summer.

**Place of Work.** It is usually on campus unless the sponsor has a special need for working on its campus, e.g., use of a special instrument or lab available only from the sponsor.

**Where to Apply.** Since the design clinic is supervised by a faculty, usually the faculty recruits the student(s) to form a team. This is another reason you need to interact more with your faculty.

**Keep a Record.** You will reap the same benefits as you do for the URE project.

## Internships

**Benefits.** Internship is a temporary on-the-job training opportunity. Because it is work experience in civil engineering, it becomes relevant to your career development. Potential employers like people with internship experiences. Some employers even hire only people who have successfully completed an internship with them. Thus a temporary job has the potential of becoming a permanent one. You will work directly under a supervisor and learn real-world knowledge.

**Types of Internship.** You may participate in a paid, unpaid, or partially paid internship. A partially paid internship pays you a stipend instead of a salary. You may work full time in summer and part time during semesters. We encourage our students to look for paid internships. Not only do you gain financially, but your employer usually takes you more seriously and gives you "real" work worthy of the money paid to you. But unpaid internship experience is also valuable if the sponsor is a reputable organization or company.

**Place of work.** Internship is usually off campus, similar to an off-campus job.

**Where to Apply.** Your faculty advisor and department chair may have information on internships. You can also search on the web for internships offered by large companies and public organizations in your area. In civil engineering, state departments of transportation and water companies usually have internship and job opportunities (see Chapter 7).

**Keep a Record.** As for Undergraduate Research Experience, you will benefit similarly to keep a record.

Many years ago internship was a part of the required curriculum for some civil engineering programs. At California State University, Northridge it is still required for one semester for the Construction Management program of the civil engineering department. This is possible because the program has a strong industrial advisory board, whose members have access to the resources needed to fund the internships for all students.

## Co-Ops

**Benefits**. Cooperative Education (Co-Op) is a system for students to alternate full-time study and full-time work but not work and study at the same time. The work is related to their major. When students work full time, they are treated as regular full-time employees and have all the benefit of a real job. It is like a try-out for a permanent job, to experience what is in the future when they graduate. When students study full time, they are no different from regular full-time students taking a full load of academic work.

**Duration**. It could be one semester of work and one semester of study or as long as a full year of work before returning to study. Rarely does it go beyond one full year of work. If students find full-time work on their own, of course they may choose to work as long as they want, but they will run the risk of giving up on their academic study altogether. Without completing their civil engineering degree, they will suffer in the long run even though they may feel rewarded at the moment working full time.

**Where to Apply**. Usually the Co-Op program is part of the civil engineering degree program and is administered by the department and supervised by faculty. Some programs also award grades.

A modified version of Co-Op, practiced at the California State University, Northridge, is called the Honors Co-Op. Only students with a GPA of 3.0 or better and are taking senior-level courses can apply. The sponsors are companies and organizations who pay students for working full-time during summer and half-time during the semesters. It is a full-year program with designated faculty and industrial representative advisors for every student. Students receive grades at the end of each semester. It has been a popular program for over 40 years and the number of participating students is limited only by the availability of industrial sponsors.

## Undergraduate Thesis and Independent Studies

Many years ago an undergraduate thesis was part of the required curriculum at most universities. Nowadays, some institutes offer it as an elective course. The demand on faculty time is significant and is similar to that of the undergraduate research experience, except a grade must be given at the end. The demand on a student's effort is also similar to that of the URE with additional commitment to write a thesis or a report (for independent studies).

## Volunteering

By volunteering your time and effort you will gain knowledge of the work and nature of what you are volunteering for. You will make an impact on other people's lives and feel a sense of accomplishment. There are numerous opportunities on and off campus. When you first come to your campus to register, you may have been greeted by one or more students who guided you. They are volunteers usually called university ambassadors. Some universities have courses that require volunteering work as part of the academic content. Off campus, almost all charity organizations need volunteers. You are encouraged to join one according to your interest or faith.

## Study Abroad

**Benefits**. By studying abroad students have the opportunity to explore another culture first hand and gaining a new perspective on their own culture. Students also learn to be independent and mature by managing their daily lives in a different world. Of course learning the use of another language is another benefit.

**Duration**. Usually studying abroad lasts for one year or shorter. Prolonged absence from the homeland is not advised.

**Where to Apply**. Many institutes now offer Studying Abroad (Education Abroad) programs that allow direct transfer of credits. The cost is reasonable and is often offset by scholarships or financial aid. There are also independent organizations specializing in study-abroad programs that you can easily find from a web search. You are encouraged to focus on your institute's program first and seek references and advice if you have to go off campus for study-abroad opportunities.

## Additional Reading Recommendations

1. *Studying Engineering: A Road Map to a Rewarding Career*, Raymond B. Landis, 4th edition, Discovery Press (www.discovery-press.com), 2013. Chapter 7 of this book is entitled "Broadening Your Education," which describes numerous co-curriculum options.

## Assignments

1. Write a one-page report listing the names of the student chapters on your campus that are mentioned in Sections 4.2 through 4.7 of this chapter.
2. The national honor society for civil engineers is called (a) Tau Beta Pi, (b) Chi Epsilon, (c) Delta Upsilon, (d) Sigma Alpha Mu, or (e) Zeta Beta Tau.
3. The national honor society for all engineers is called (a) Tau Beta Pi, (b) Chi Epsilon, (c) Delta Upsilon, (d) Sigma Alpha Mu, or (e) Zeta Beta Tau.
4. The oldest engineering society in America is (a) ASCE, (b) ASME, (c) IEEE, (d) AWWA, or (e) SWE.
5. The Annual National Concrete Canoe Competition is sponsored by (a) American Concrete Institute (ACI), (b) ASCE, (c) AISC, (d) ASCE/AISC jointly, or (e) ACI/ASCE jointly.
6. The Annual Steel Bridge Competition is sponsored by (a) American Concrete Institute (ACI), (b) ASCE, (c) AISC, (d) ASCE/AISC jointly, or (e) ACI/ASCE jointly.
7. Each student chapter of ASCE has one faculty advisor and (a) 1, (b) 2, (c) 3, (d) 4, or (e) zero Practitioner Advisors.
8. The idea of the National Concrete Canoe Competition was initiated by a (a) professor, (b) practitioner, (c) contractor, (d) concrete company, or (e) none of the above.
9. The concrete used in the concrete canoe competition is (a) designed by ASCE, (b) designed by the competing teams, (c) prepackaged concrete, (d) premixed concrete, or (e) none of the above.
10. Among the following five national engineering organizations, the oldest is (a) AWWA, (b) ACI, (c) SWE, (d) NSBE, or (e) SHPE.
11. Co-Op means (a) Co-Ed, (b) Corporations for Education, (c) working part time while studying full time, (d) working part time while studying part time, or (e) alternating full-time work and full-time study.
12. In the National Steel Bridge Competition the award categories include the (a) construction speed, (b) construction economy, (c) lightness, (d) stiffness, or (e) all of the above.

# Chapter 5

# Legends, Milestones, and Landmarks

## 5.1 Overview

As civil engineering has always dealt with the basic needs of everyday life such as shelter and water, its history goes back to ancient times. Early designers and builders practiced trial and error and used experience accumulated in generations to achieve greatness, all without the benefit of modern science and technology. As described in this book, modern civil engineering is deeply rooted in the application of mathematics and science. While the genesis of mathematics and science can be traced back more than four thousand years, the main foundation for civil engineering is mechanics, which was developed only in the last five hundred years. Furthermore, civil engineering as practiced today is as young as the advent of the modern digital computer. In this chapter, some of the most famous landmarks and the most important people in the development of civil engineering, from ancient times to modern times, are described. Brevity necessitates the very selective inclusion of only a few American landmarks and legends when it comes to modern times. Ancient achievements from other cultures are included only for the most significant. Pictures of landmarks such as the Great Pyramid and all the others included herein are readily available on the internet and only a few are reproduced herein.

## 5.2 Ancient Monuments and Landmarks

**The Great Pyramid**. Of the Seven Wonders of the Ancient World around the Mediterranean rim described by Greeks 2,200 years ago (Great Pyramid of Giza, Hanging Gardens of Babylon, Temple of Artemis at Ephesus, Statue of Zeus at Olympia, Mausoleum of Maussollos, Colossus of Rhodes, and the Lighthouse of Alexandria), only the Great Pyramid of Giza in Egypt still stands today. Of the 100 or so pyramids built by the ancient Egyptians, the Great Pyramid of Giza is the grandest and tallest (146.6 m or 481 ft high). It is also called Khufu's Pyramid because it is the tomb of the Fourth Dynasty Egyptian King Khufu (2551–2528 B.C.).

Built around 2540 B.C., it is also the oldest and largest of the three pyramids at Giza near present-day Cairo, Egypt. Khufu's son Djedefre succeeded Khufu as pharaoh but reigned only eight years and was succeeded by another son of Khufu, Khafre, who built the second major pyramid at Giza. The Khafre Pyramid is not as tall as the Khufu Pyramid with a height of 143.5 m (470.8 ft). Khafre's son Menkaure built the last and the smallest of the three pyramids at Giza with a height of only 65 m (213.5 ft). All three pyramids at Giza are of similar design and construction and are a major departure from and improvement on previous pyramids.

Many scholars have studied the design and construction of the Egyptian pyramids. The latest book and the most comprehensive, published in 2004, entitled *How the Great Pyramid Was Built*, authored by Dr. Craig B. Smith, provides numerous new findings. Dr. Smith is an experienced engineer, a construction executive, and former president of a major global engineering, architecture, and construction company, in addition to being a well-known scholar on pyramid construction. His book provides the main source for much of the descriptions herein.

The evolution of pyramid design and construction can be summarized in the three different types of pyramids preceding the Khufu Pyramid: The Step Pyramid (2620 B.C.), The Bent Pyramid (2565 B.C.), and the Red Pyramid (2560 B.C.), as shown in Figure 5.1.

The step pyramid is named for its shape. At a height of 60 m, it presented an advance in building stone structures to greater heights at the time. It was built for the Third Dynasty pharaoh Djoser, who reigned from 2635 to 2610 B.C., by his chief architect Imhotep. The Bent Pyramid was built by the first pharaoh of the Fourth Dynasty, Sneferu. He reigned from 2613 to 2589 B.C. It was built on sandy desert soil, the bearing capacity of which was not enough to support the originally designed mass of the pyramid with an inclination angle of 54.5 degrees. Subsidence was observed during initial construction. Consequently approximately one-third of the way up, the inclination is bent inward to an angle of 43 degrees, thus the name Bent Pyramid. The Red Pyramid, named for its red-pink limestone, was built by the same pharaoh Sneferu who apparently was seeking an improved new pyramid from the Bent Pyramid. It was built on a better foundation. The builders placed the king's burial chamber above the ground and in the body of the pyramid for the first time. The success of the Red Pyramid paved the way for the Great Khufu Pyramid, built by Sneferu's son Khufu. The Khufu Pyramid was built on a leveled strong rock base at a new site outside today's Cairo. Its King's Chamber and Queen's Chamber are located high in the body of the pyramid.

The present-day Khufu Pyramid has a height of 138.8 m (455 ft), which does not represent its original height because of erosion and its missing capstone, the pyramidion. The measurements done in 1880–1882 by the English Egyptologist Flinders Petrie (1853–1942) and later studies published in his book *The Pyramids and Temples of Gizeh*, in 1883, point to an original dimension of 280 cubits in height and 440

Figure 5.1 The Step (Left), Bent (Middle), Red (Right) pyramids.

Figure 5.2 The Great Khufu Pyramid.



cubits in length at each of the four sides of its base. The cubit is an ancient Egyptian length unit that may have inspired the later English unit "foot," because a cubit was defined as the length of pharaoh's arm measured from his armpit to the farthest tip of his fingers. Numerous cubit rods were excavated and it was determined at Khufu's time that one cubit is equivalent to 0.524 m (20.6 in). Thus the original height of the Khufu Pyramid was determined as 146.6 m (481 ft) and its base was 230.4 m (756 ft) and its side faces inclined with an angle of 51.9 degrees. It is interesting to note that the base-to-height ratio of 440/280 = 1.571 is identical to $\pi/2=1.571$. Is this a mathematical coincidence or by design? Another mathematical coincidence/by-design pointed out by the Greek historian Herodotus in 450 B.C. is that the square of the height of the pyramid is equal to the area of each of the inclined faces of the four triangles. The enormity of the size of the base can be put in perspective by comparison to the standard football field in America: The size of the base of the Khufu Pyramid is close to the area of ten football fields. The height of 146.6 m held the world's height record for 4,300 years until the erection of the 169 m-high Washington Monument in 1885. Another interesting feature is the heights of the 218 courses (layers) are not constant. At the base, the height is 1.49 m (59 in). It generally becomes smaller going up, though it spikes up at irregular levels and then trends down. At the highest level it is less than 0.5 m (19.7 in).

Dr. Craig Smith described survey and measuring tools developed by the ancient Egyptians. Using the tools of their times, the builders of Khufu Pyramid achieved astounding precision. The four corner angles of the square base are off from a perfect 90 degrees by no more than 0.06 degrees. In fact two of the four corner angles are equal to a perfect 90 degrees. The four base lengths deviate from the average of 230.4 m (756 ft) by no more than 0.11 m (4.3 in).

The number of stone blocks is 2.2 million for internal core stones and 98,000 for casing stones, the exterior face stones. The weight of the internal blocks ranges from 1,359 kg to 14,656 kg (1 kg = 2.2 lb) according to Dr. Smith's computation. The transportation and placement of these heavy blocks and the time needed to complete the project has been the subject of numerous studies and speculations. Some even ventured the theory that extraterrestrial visitors must have been the source of advanced technology unknown to the Egyptians at the time. Considering all the available tools and knowledge of the Egyptians, Dr. Smith convincingly concluded that a multi-ramp approach in combination with tools to move large weights explains the way to raise the stone blocks to their spots in the pyramid. He also estimated that the project required a permanent work force of 4,655 including direct labor such as stonemasons, brick-makers, carpenters, foundry men, tool sharpeners, surveyors, rope makers, painters/artists, sculptors, and indirect labor such as supervisors, overseers, scribes and clerks, cooks, bakers, brewers, stevedores, warehouse workers, doctors and priests, and security were needed at the time, which was definitely achievable by the Egyptians. A detailed construction schedule, including all phases of a construction project worked out by Dr. Smith, shows that the Khufu Pyramid could have been completed in about ten years. His "optimal" scenario would require, at its peak demand of laborers, an additional 21,000 temporary workers to assist the quarry workers, and move stones to the site and onto the pyramid.

The Khufu Pyramid was covered with white limestone casing blocks with inclined outer surfaces when it was finished. Today one can only imagine what a magnificent sight a shining white monument, standing above the desert horizon, must have been; only the top one-third still has the casing stones mostly in place.

While Khufu was the king who commissioned the Great Pyramid, the man most responsible for the design and construction of the pyramid was Hemiunu, a cousin of Khufu. His title, roughly translated, was Master of Work and Vizier. His importance can be indicated by the site of his own tomb. Discovered in 1912, his tomb was located in the west cemetery behind the pyramid and one of the largest in the cemetery. His tomb contained a life-sized sitting stone statue of him. The statue is now in the Roemer and Pelizaeus Museum in Hildesheim, Germany.

**The Great Wall of China**. While the Egyptian Pyramids were built to honor the dead rulers, the Great Wall of China was built for a more practical purpose: to defend against the invasion of northern nomadic tribes, although its effectiveness was marred by the frequent success of the northern invaders in crossing the wall throughout the ages ever since the wall first appeared around or before the fifth century B.C. There

Figure 5.3 Entrance (Left), Corbel-Arched Gallery (Middle), and Statue of Hemiunu (Right).

Figure 5.4 The Great Wall of China (Left), The East End (Right) at Shanhai Pass.

Source: http://en.wikipedia.org/wiki/File:Great_Wall_of_China_July_2006.JPG. Copyright in the Public Domain.
Copyright © Fabien Dany (www.fabiendany.com), (CC BY-SA 2.5) at http://en.wikipedia.org/wiki/File:Chemin_de_ronde_muraille_long.JPG.

is a famous story about a tyrant around that time amusing his favorite concubine by falsely raising fire on the watchtowers to summon the troops to the capital. The watchtowers are part of the great wall system, which consists of long stretches of earthen or brick walls dotted by the towers. When northern invaders were spotted, the defenders would fire up the camel dung from the watchtowers. Legend has it that the smoke from camel dung could be seen from afar. One tower after another signaled the invasion until the news reached the central government for urgent action. That was the defense system set up by the ancient Chinese. The Great Wall is one of the Seven Wonders of the World and was declared a UNESCO World Heritage site in 1987.

History books also described the human sacrifices in building the wall, especially during the reign of the First Emperor of China around 221 B.C. when he unified the Warring States into a single central government, the Qin dynasty. The present-day name of "China" was derived from "Qin." He destroyed several existing walls separating the formal warring states and built a new northern wall. Later dynasties built new walls connecting existing walls. There are large local loops of walls, presumably for the defense of local areas, connected to the long stretch of walls. The present-day Great Wall is believed to have been completed during the Ming dynasty around the year 1450 A.D. It stretches from the eastern seaboard of China just northeast of the present-day capital, Beijing, to the west, through Inner Mongolia, and ends in the western reaches of the Gansu province.

The wall was "great" because of its extraordinary length. Ironically the question of how long the great wall is does not have a definitive answer. According to the famed scholar on Chinese Science and Technology, Joseph Needham (*Science and Civilization of China*, Volume 4, Part 3. Civil Engineering and Nautics, Cambridge University Press, 1971), the main line of the Great Wall has a length of 2,150 miles (3,060 km) and the loops have a length of 1,780 miles (2,864 km), adding up to 3,930 miles (5,944 km). Since Needham's time, new discoveries and measurements (using GPS) were made. Today, the total length of the Great Wall is reported as 5,500 miles (8,851 km). See *Sydney Morning Herald*, April 20, 2009 (www.smh.com.au). But in 2012, a five-year archaeological survey, done by the Chinese State Administration of Cultural Heritage (SACH), found that the total length of the Great Wall was 13,170 miles and reached across 15 provinces (http://abcnews.go.com/blogs/headlines/2012/07/great-wall-of-china-longer-than-previously-reported/).

The body of the wall is made mainly from compacted earth. The walls built during the Ming Dynasty are stronger with the generous use of bricks and stones at its base and at the outer layers. A typical section of the wall has a height from 6 to 10 m (20 to 30 ft) and a base width of 8 m (25 ft). The watchtowers are wider (roughly 50% more) and taller (roughly 40%–60% more).

**The Roman Aqueducts.** The Romans built aqueducts throughout its empire to supply water to its cities. An aqueduct is a channel/bridge that was built to convey water from one place to another. The most famous are the eleven aqueducts built from 312 B.C. to 226 A.D. for the City of Rome with a total length of 418 km (260 miles). The longest was 95 km (59 miles). The construction of these aqueducts required careful planning and engineering. When the water reached the city, it was temporarily gathered at large pools and distributed to the emperor, sold to rich citizens, and provided free to the public through the numerous fountains. The water was not stored. Any excess was used to flush out sewers.

As only gravity was used to drive the water, the aqueducts were built with a gentle grade of about 1/200 throughout their entire lengths. Although pictures of Roman aqueducts often show magnificent arch construction, less than 10 percent, 47 km (30 miles), of Rome's 418 km aqueduct system crossed over valleys on raised stone arches. Most of the rest ran through underground conduits made mostly of stone and terra cotta pipes. The above-ground channel itself was about 1 meter wide (3 ft.) and 2 m high, allowing a man to walk through. It is estimated that the water supplied to Rome on daily basis was sufficient to provide for a population of one million. Such a vast aqueduct system requires constant maintenance. The Roman had an appointed official, Curator Aquarum, to oversee the maintenance.

The Roman aqueduct system was the grandest in ancient times, but it was not the first. Similar systems were built earlier in ancient Persia, India, Egypt, and other Middle Eastern countries. While the Roman aqueduct began to deteriorate around the 4th and 5th centuries, other similar systems were built throughout that time. Modern aqueducts are longer in length and larger. New York City uses three main aqueduct systems to supply 1,800,000,000 gal (6,800,000,000 liters) of water a day from sources 190 km (120 miles) away. The largest aqueduct system in the world is the famous California Aqueducts.

**California Aqueducts.** The aqueduct system in the state of California is by far the largest in the world. There are three main systems: Owens River Aqueduct (LA Aqueduct), Colorado River Aqueduct, and the

Figure 5.5 Tarragona Aqueduct, Spain (Left), Tarragona Aqueduct, Spain view from the top (Right).

Figure 5.6 Water entering the LA Aqueduct (Upper Left); LA Aqueduct Cascade (Upper Right);Parker Dam, Arizona of the Colorado Aqueduct (Lower).



Governor Edmund G. Brown California Aqueduct. They were designed to provide water for municipal and agricultural uses in Southern California.

**Owens River Aqueduct**: Also referred to as Los Angeles Aqueduct, the Owens River Aqueduct transports water from the central valley of California to the City of Los Angeles. The project started in 1907 and its initial phase was finished in 1913. The aqueduct ran 373 km (232 miles) from an intake at Independence, CA, through mountainous desert to Owensmouth in the San Fernando Valley, the northern part of the City of Los Angeles. It includes 24 miles of unlined canal, 37 miles of lined canal, 98 miles of pipe conduit, 43 miles of lined tunnel, and 12 miles of steel inverted siphons. Since 1913 the aqueduct has been extended northward another 100 miles to reach additional sources of supply. (See *Historic Civil Engineering Landmarks of Southern California*, by William Myers, The History and Heritage Committee, Los Angeles Section, ASCE, 1974.)

**Colorado River Aqueduct**: The work for Colorado River aqueduct started in 1932 and was completed in 1941. Its 389 km length (242 miles) runs from the Parker Dam near Lake Havasu City, Arizona, to Lake Mathews in western Riverside County of Southern California, through some of the worst desert and mountainous terrain. It has 5 pumping plants and delivery lines, 148 km (92 miles) of tunnels, 101 km (63.9 miles) of canal, 88 km (54 miles) of cut-and-cover conduits, and 44 km (27 miles) of inverted siphons.

The Owens River Aqueduct and the Colorado River Aqueduct provide the water consumed by the City of Los Angeles and the surrounding areas not only for municipal uses but also for agricultural uses. The man most responsible for the two aqueducts was **William Mulholland** (1855–1935). Mulholland was born in Belfast, Ireland. He became a sailor and came to America in the 1870s, and worked in various jobs before moving to the Los Angeles area. In 1887 he began his work as a ditch cleaner for the private water company of Los Angeles. He rose through the ranks to become the superintendent of the company in eight years. When Los Angeles took over the company in 1902 and formed the Department of Water and Power, he was retained as its head, a job he held until 1928. During his tenure, the Owens River Aqueduct was built, not without political maneuver and deceit. The controversial story of his political endeavors was used as the background for the 1974 Oscar-winning movie *Chinatown* (for Best Writing, Original Screenplay). In 1923, he recommended a survey to investigate the feasibility of transporting water from the Colorado River to Southern California. That investigation eventually led to the construction of the Colorado River Aqueduct. Today, a major street in Los Angeles is named after him, so are a middle school and the William Mulholland Memorial Fountain in Los Angeles.

**California Aqueduct (Governor Edmund G. Brown California Aqueduct)**: Started in 1960, the 714 km (444 miles) aqueduct is the world's largest. The aqueduct begins at the San Joaquin–Sacramento River Delta in northern California and flows south. It has three branches serving the central coast, central valley, and southern California. A typical section is concrete lined, 12 m (40 ft) wide at the base, with an average depth of flow of 9 m (30 ft). The system has 20 pumping stations, 130 hydroelectric plants, and more than 100 dams and flow-control structures.

**Du-jiang-yan Water Conservancy Project**. The Du-jiang-yan Water Conservancy Project consists of a system of hydraulic engineering constructions initially built around 256 B.C. in today's Sichuan Province of China, on the Min River upstream from the provincial capital of Chengdu. It is still in use today providing water for irrigation and municipal consumptions for an area of 6,687 $km^2$ (1.6 million acres) in 40 counties, roughly the size of the state of Delaware. It is the world's oldest and largest water conservancy project.

In ancient China, the Min River, a large branch on the upper reaches of the Yangtze River, was an annual flood hazard for the people living on the Chengdu Plain. By 256 B.C. during the Warring States period of China, the Kingdom of Qin allocated funding to the Head of the Sichuan Province, Li Bin, to solve the irrigation and flooding problem. Li and his son, working with local farmers, completed three major projects (Treasure Bottle Pass, Fish Mouth Levee, and Flying Sand Weir) in 14 years that forever eliminated the annual flooding problem and made Sichuan the "rice warehouse" of the region for more than 2,200 years. The three projects Li and his son completed achieved automatic river flow diversion, automatic flow control, automatic flushing, and flood elimination. The father-and-son team established easy-to-follow rules to keep the irrigation channels downstream open and functional. These rules were followed through the ages and even today. Thus, the Li team achieved this great water conservancy project over two thousand years ago by innovative design, construction, and operation/maintenance very much in sync with modern practices.

The Du-jiang-yan project was a tourist attraction throughout the ages. During the reign of the Mongolian Emperor Kublei Khan (1264–1294 A.D.) of the Yuan Dynasty, the famed Italian traveler Marco Polo visited the project site by traveling on horseback for twenty days from the northern province of Shaanxi. In his book *Travels of Marco Polo*, he wrote "The Du-jiang river system has rapid flow, abundance of fish, many boats carrying commerce go up and downstream."

Figure 5.7 The California Aqueduct.



The May 12, 2008 Wenchuan Earthquake of magnitude 7.9 struck the Du-jiang-yan area but incurred only minor damage to the project. In fact the project was among the first to reopen to tourists.

**The Grand Canal of China.** The world's oldest and longest canal is the Grand Canal of China. It is also called the Jing-Hang Grand Canal because it connects Beijing in the north and Hangzhou in the south. Its total length is 1,774 km (1,102 miles). The beginning of the canal can be traced to more than 2,600 years ago but it was completed to its present-day length about 800 years ago. The main work of the canal was credited to Emperor Yang of the Sui dynasty, who started the work around 600 A.D. He was also blamed for the excessive taxing of his citizens to support the canal project.

The canal is connected to five major river systems of China, including the Yellow River and the Yangtze River. Joseph Needham's study showed that at the end of the 19th century, there were 13 sections in the canal system with water depth varying from 3 m (10 ft) to 16 m (49 ft) (*Science and Civilization of China*, Volume 4, Part 3. Civil Engineering and Nautics, Cambridge University Press, 1971).

For centuries, the canal was the main conduit to ship rice and goods from the abundant south to the political center in the north. The annual tribute (tax) of rice directly supported the imperial court and its officials. The importance of the canal waned in the late 19th century with the opening of railways and sea lanes.

Today, the canal is integrated with the other river systems of China to provide irrigation and for commerce shipping. Ships with tonnage of 500 or below can navigate the southern 660 km yearlong. Seasonal

Figure 5.8 The Du-jiang-yan Weir (Left) and the Fish Mouth Levee (Right).



navigation extends to 1,100 km. Ongoing improvement will allow 1,000-ton ships to navigate the southern 60% of the canal by 2015.

**Ancient Arch Bridges**. The Romans were credited for building arch bridges, although corbel-arch bridges more than 3,000 years old were found in Greece. There were many different types of ancient arch structures. We shall explore only three.

The corbel arch is the easiest to construct. The weight of the construction and the load introduces tension in the arch, limiting its applications to creating narrow passageway underneath or to extend over small spans. It has been in use since Egyptians built their pyramids.

Romans used semicircular arches, in which the load was transmitted to the abutments through compression. Its span is limited to the radius of the semicircle.

Figure 5.9 Modern Course of the Grand Canal (Left) and one southern section (Right).

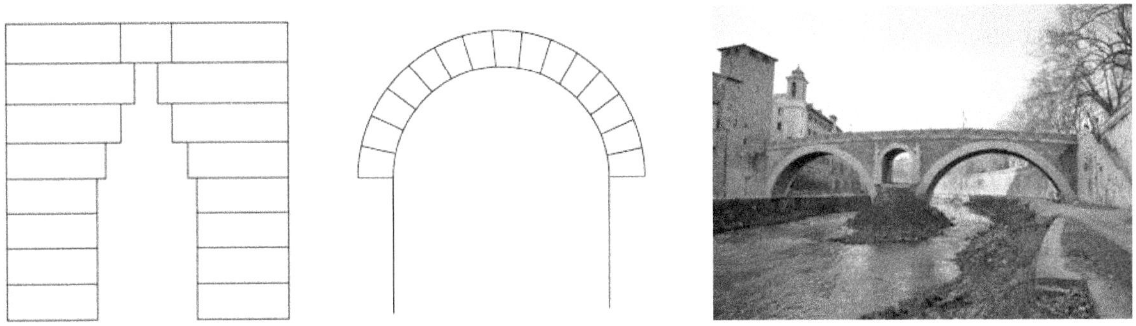

Figure 5.10 Corbel Arch (Left), Round Arch (Middle), and the Pons Fabricius Bridge in Rome (Right).



To widen the span, the segmental arch, using part of the semicircle, was created. When the span-to-rise ratio becomes larger, however, more regions in the arch would be under tension. Thus, the art of construction is to balance the need for larger span and the need for limiting tension. The Pons Fabricius Bridge in Rome, Italy, is the world's oldest segmental arch bridge still standing today. It has two nearly identical spans, with the larger span at 24.5 m (80 ft). It was built in 62 B.C. Although it is segmental, it is very close to semicircular with a span-to-rise ratio of 2.4.

The Zhaozhou Bridge is the world's first stone open-spandrel segmental arch bridges, built around 610 A.D. in northern China. It has a span of 37.31 m (122.4 ft) and a rise of 7.23 m (23.7 ft), resulting in a span-to-rise ratio of 5.12:1 (*History of Ancient Chinese Bridge Technology*, by Yisheng Mao, Ming Wen Book Co., Ltd, Taipei, Taiwan, 1991, In Chinese). The main span is built with 28 pieces of limestone narrow blocks connected by iron dovetails. On each side there are two open smaller arches. Because of this design the bridge has survived numerous natural disasters such as flooding and earthquakes for more than 1,400 years. It is now a tourist attraction and a protected cultural heritage site. After Zhaozhou Bridge, the spandrel-arch design appeared in many 12th-century Chinese bridges, but it did not appear in Europe until the 14th century. The Zhaozhou Bridge stands today as one of the international landmarks recognized by the American Society of Civil Engineering.

## 5.3 Modern Landmarks

**The Brooklyn Bridge**. When the Brooklyn Bridge, spanning the East River between Manhattan and Brooklyn, was completed in 1883, it was the longest bridge in the world, the first in using pneumatic caissons to build its foundation, and the world's first steel cable suspension bridge. Its 486 m (1596 ft) span was barely surpassed in 1903 by the nearby Williamsburg Bridge (488 m/1600 ft), also connecting Manhattan and Brooklyn.

The engineering achievements of the Brooklyn Bridge cannot be told without mentioning the human story of the three people most responsible for its design and construction: John, Washington, and Emily Roebling. **John Augustus Roebling** (1806 –1869) was educated as a civil engineer at the Royal Polytechnic

Figure 5.11 The Zhaozhou Bridge in China.

Source: http://en.wikipedia.org/wiki/File:Zhaozhou_Bridge.jpg. Copyright in the Public Domain.

School of Berlin, Prussia (later Germany). He immigrated to the new world in 1831 and bought a tract of land west of Pittsburg, Pennsylvania, and started a farm community that he named Saxonburg. In 1837 he was appointed the engineer of the Pennsylvania Canal project. He developed steel-wire-rope technology and found its first application at the Allegheny portage canal project in 1842. By 1844, the John A. Roebling Company built a suspension aqueduct across the Allegheny River. The Roebling wire rope was a huge success and was used in the first Otis elevators in 1862. Between 1844 and 1957, Roebling built five suspension bridges in Pennsylvania and New York using his wire ropes. Legend has it that in 1852 when John and his eldest son **Washington**, then 15, were waiting for the ferry to take them from Brooklyn to Manhattan, his vision of a suspension bridge crossing the East River was formed.

Brooklyn was then a separate city. The only means of crossing the East River was through ferry, the operation of which was weather dependent and dangerous. Roebling's campaign for a bridge was approved in 1866 when the New York State Legislature passed a bill for the construction of the Brooklyn Bridge. John Roebling was appointed chief engineer of the Brooklyn Bridge in 1867. Roebling's design of the bridge was mostly completed in 1869. One day in July an incident at the Fulton ferry slip in Brooklyn crushed one of his feet.

Figure 5.12 Brooklyn Bridge viewed from Brooklyn, with downtown Manhattan as a backdrop.

Source: http://en.wikipedia.org/wiki/File:Brooklyn_Br%C3%BCcke.jpg. Copyright in the Public Domain.

The injury and the ensuing infection claimed his life after only 16 days. The Brooklyn Bridge project was now in jeopardy, having lost its designer and chief engineer. His son Washington Roebling came to the rescue.

Washington Roebling got his civil engineering degree from the Rensselaer Polytechnic Institute in Troy, New York. He assisted his father in several suspension bridge projects before joining the army in 1960. During the Civil War, he participated in the Battle of Gettysburg. After the war he worked with his father on the Cincinnati-Covington Bridge (now the John A. Roebling Suspension Bridge). This experience turned out to be critical to his later responsibility with the Brooklyn Bridge. By 1868 he was appointed assistant engineer of the Brooklyn Bridge project. He was appointed chief engineer 17 days after his father's death, at the age of 32.

Washington immediately went to work and finalized the details of the pneumatic caisson. The construction began in January, 1870, and the Brooklyn-side caisson was launched in March. The wooden caisson was a huge rectangular (168 ft x 102 ft) inverted box of 14.5 ft depth. Its top was 15 ft thick and made of yellow pine. (See *The Builders of the Bridge: The Story of John Roebling and His Son*, by D.B Steinman; Harcourt, Brace and Co., Inc., 1945.) As the caisson was lowered, the tower of the bridge was constructed layer by layer. Workers went down to the river bed through two pressurized access shafts and excavated the river bed rock and soil to be raised to the top through two water shafts. The Brooklyn caisson reached the solid-enough bed at 44.5 ft depth and was completed in March, 1871 after its launch a year earlier. The New York-side caisson was launched in May, 1871. Although it was also completed in one year, its progress was agonizing for Washington. It went down without reaching any significant support until 78 ft below the water level. After carefully investigate the soil strength, Washington decided the bearing capacity was sufficient and there was no need to go further. His decision turned out to be correct, but at the time it also had a lot to do with the disease that workers suffered and he also suffered. Because the workers were in a pressurized environment for extended time and did not know at the time that they needed to come up slowly, they suffered the diver's disease, called "the bends" then. Several died. Washington spent more time in the caisson than any of his assistants and in 1872, he was so ill that he no longer had the strength to go to the site. Again the Brooklyn Bridge project was in jeopardy. His wife, **Emily**, came to the rescue. From that point on, he directed the construction from his residence in Brooklyn through Emily. Although not an engineer by training, Emily, through sheer determination and intelligence, was able to communicate effectively with the construction team to carry out Washington's directives in every detail.

The Brooklyn and New York towers were completed in 1875 and 1876, respectively. In August 1876, the first wire was passed between the two towers. The Brooklyn Bridge has four main cables, each with a diameter of 15.75 in., consisting of 19 strands. Each strand in turn contains 286 steel wires. Washington, in order to strengthen the bridge, added 400 diagonal stays, connecting the bridge deck structure to the two towers. The diagonal stays and the 1,520 suspenders (connecting the cables to the deck structure) create a unique appearance of the bridge. (See *The Brooklyn Bridge: They Said It Couldn't Be Built*, by Judith St. George, G.P. Putnam's and Sons, New York; 1982.) The manufacturing and placing of the cables and the construction of the bridge deck would take seven more years. When it was opened in May 1883, 14 years had passed since the construction began. On opening day, President Chester Arthur attended the ceremony and Emily Roebling was given the honor of being the first to cross the bridge by carriage.

The Brooklyn Bridge is still the world's 73rd-longest bridge and a major tourist attraction. The Brooklyn Bridge became a symbol of the triumph of the human spirit.

**The Golden Gate Bridge**. The record of the world's longest span bridge was broken in 1937 when the 4,200 ft (1,280 m) span Golden Gate Bridge was completed.

The need for a bridge to cross the Golden Gate between the northern tip of the San Francisco Peninsula and Marin County to the north was similar to that for the Brooklyn Bridge. Ferry was the only means of crossing the bay. By 1928 over 2 million automobiles used the ferry service. The overcrowding reached a crisis when traffic from Sausalito at the north had to wait until Tuesday morning to return to the city following a long Labor Day weekend.

The call for a bridge can be traced back to 1869 but little attention was given to the idea until 1916 when a newspaper reporter and a former engineering student by the name of James Wilkins began a campaign for the bridge. His effort eventually prompted the city engineer Michael O'Shaughnessy to solicit design proposals from three engineers in 1920. Only Joseph Strauss responded with a design involving a combination of a cantilever bridge and suspension bridge. Strauss was born in Cincinnati, Ohio, in 1870 and raised in the shadow of a suspension bridge in Cincinnati designed by John Roebling. At 1,057 ft (322 m) it was proclaimed to be the longest bridge when it was opened in 1866. Legend has it that this bridge by Roebling inspired Strauss to be a bridge engineer. He graduated from the University of Cincinnati and entered a career as a bridge designer. He was credited for the design of many draw bridges. Strauss eventually became the chief engineer for the Golden Gate Bridge.

By 1930 Strauss's original design was changed to involve only a single suspension bridge. The structural design was credited to his senior engineer **Charles Ellis**. Ellis was a maverick, and taught engineering at University of Illinois before he obtained a civil engineering degree at the university. No other bridge's construction was more celebrated than the Golden Gate. When its official groundbreaking took place in February 26, 1933, a parade was held and a congratulatory telegram from President Hoover was read. The two towers were constructed using steel members riveted together and its height of 746 ft (224.7 m) is still

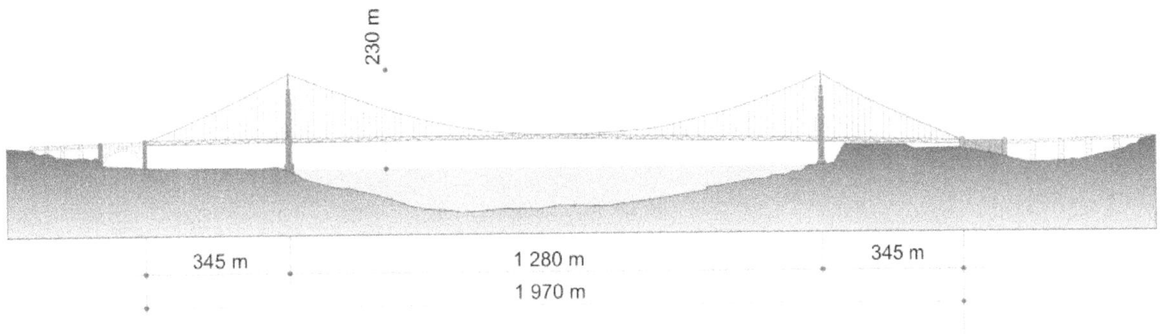


Figure 5.13 The Golden Gate Bridge.



the highest in U.S. among bridge towers. Two steel cables, each with a diameter of 35 in. (55 cm), span over the bridge. Each cable has 61 strands and each strand has 452 steel wires. Thanks to Strauss's invention of putting a running safety net under the cables, the construction had the best safety record in bridge construction with only eleven deaths; ten of whom fell onto the net and broke through.

The opening of the bridge was no less ceremonial, with two official opening days. May 27, 1937 was for pedestrians and May 28 was for automobiles. The Golden Gate Bridge has another first: the first to institute a toll system that collects tolls only one way. This does not reduce the revenue because it simply collects double the amount. The Golden Gate Bridge still ranks 9th in the world for its long span, but its fame surpasses all the other bridges.

**The Verrazano-Narrows Bridge**. Perhaps one of the least-known major suspension bridges in America is the Verrazano-Narrows Bridge, despite the fact that it is still the longest suspension bridge in America and ranked 8th in the world with its 4,260 foot (1,298 m) span. The bridge crosses the Narrows that separate Upper New York Bay from Lower New York Bay at the south end of the Hudson River, and connects Staten Island to Brooklyn. It was named after the Italian explorer Giovanni da Verrazano (1485–1528), who was the first European navigator to enter New York Harbor and the Hudson River.

When it was completed in 1964, the Verrazano-Narrows Bridge took away the world's longest suspension bridge title from the Golden Gate Bridge. It was the work of the famed Swiss-American bridge designer **Othmar Ammann** (1879–1965). Educated at the Federal Technological Institute in Zurich (Switzerland), Ammann immigrated to the U.S. in 1904 and worked on bridge design. Among his great achievements in bridge design are the George Washington Bridge and the Verrazano-Narrows Bridge. His bridges are known for their simplicity in shape. In 1946, he and Charles S. Whitney co-founded Ammann and Whitney, a design firm that to this day still specializes in bridge design and rehabilitation. Another legendary name associated with the Verrazano-Narrows Bridge is **Robert Moses** (1888–1981). As the New York State Parks Commissioner and head of the Triborough Bridge and Tunnel Authority, Moses oversaw the Verrazano-Narrows Bridge project (*Robert Moses and the Modern City: The Transformation of New York*, edited by Hilary Ballon and Kenneth T. Jackson). Moses was credited as the "Master Builder" for New York during the 1930–1950 period.

**The Empire State Building**. When the 102-story Empire State Building was opened on May 1, 1931, it was the world's tallest structure. The roof ends at a height of 1,250 ft (381 m) but the mooring mast/antenna extends to a height of 1,472 ft (448.7 m).

Figure 5.14 The Golden Gate Bridge viewed from the Marin Headlands.



From the outset the Empire State Building was meant to be the world's tallest. It was the brainchild of Al Smith and John Raskob. Born to Irish immigrant parents in 1873 in the shadow of the Brooklyn Bridge, Smith began working odd jobs at the age of 15. At 22 he was appointed as a subpoena server for the commissioner of jurors. He was elected to the New York State Assembly at 30. He became a reformer in political circles and fought for social legislation such as workers' compensation and limiting working hours of women and children. He was elected governor of New York in 1918. He was a flamboyant and charismatic politician with an amazing memory and grasp of details. His tenure as governor coincided with the prosperous years after the end of World War I. In 1928 he was the Democratic Party's nominee for president but lost the campaign to the more conservative Herbert Hoover. After the defeat, he teamed up with John Raskob and launched the Empire State Building project. Raskob had similar humble beginnings to Smith. Born in 1879 in Lockport, New York, he worked as a bookkeeper for Pierre S. du Pont at the age of 21. Three years later, he became the assistant to Pierre when Pierre became the treasurer of the E.I. du Pont de Nemours & Co. He had the vision that the automobile industry would be important and persuaded du Pont to invest in General Motors. Soon du Pont became a major stock holder and GM chairman, and

Figure 5.15 Verrazano-Narrows Bridge at Night from Brooklyn.



Raskob the vice president and chairman of the finance committee of GM. He was the man who created the General Motors Acceptance Corporation (GMAC), providing credit to buyers. How he teamed up with Smith on the Empire State Building project was a matter of various legends. (*Empire State Building: The Making of a Landmark*, by John Tauranac, published by Scribner, 1995.)

On August 29, 1929 the Empire State Building project was made public. By December, Smith announced that the building would be 1,250 ft high, more than 202 ft higher than the Chrysler Building, then still under construction. Furthermore, the opening date was set as May 1, 1931, a date chosen for commercial considerations. May 1 is the traditional starting date for renting contracts. To meet the deadline, the architecture firm of Shreve, Lamb and Harman and the contractor Starrett Brothers and Eken worked swiftly to start design and construction planning. The resulting design was a simple, elegant, slender building with steel girders and columns. The site was in lower Manhattan where the bedrock was strong enough to support a massive high-rise. At the basement level, 210 steel columns rose upward from simple footings sitting directly on the bedrock. Twelve of the 210 columns would rise all the way to the top, while others were terminated at different levels where moderate setbacks were made.

Careful planning by the designer and contractor resulted in an amazing speed of construction. The girders and column sections were made to one-eighth-inch accuracy at the American Bridge and McClintock-Marshall plant in Pittsburgh, shipped to a staging yard in Bayonne, New Jersey, to be marked for its final

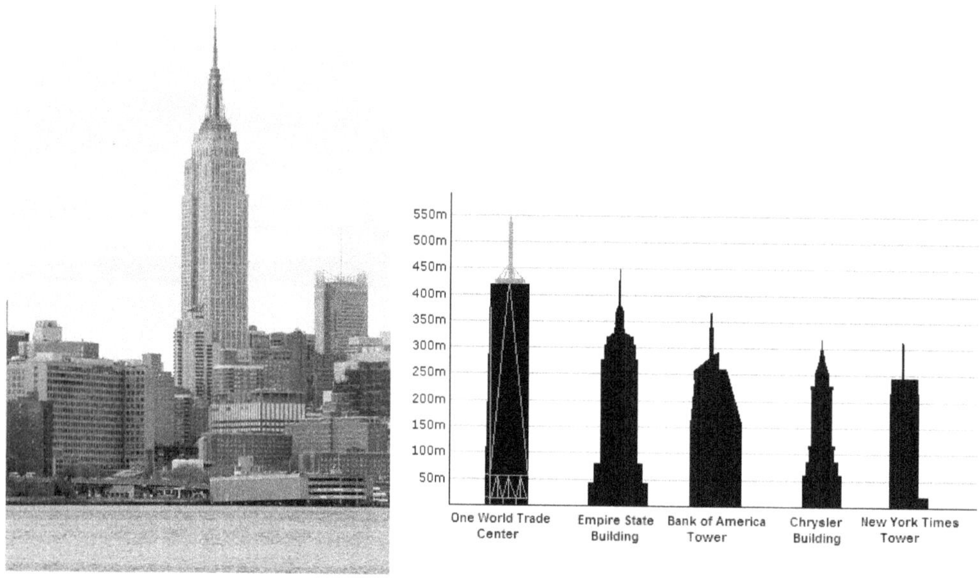


Figure 5.16 The Empire State Building seen from Brooklyn (Left), Height comparison of New York City Buildings (Right).



location in the building, and then sent to the construction site to be immediately erected. The whole process took only 80 hours for most sections. The resulting speed was four and a half stories per week while workers worked only five days a week and seven and half hours a day. By November 1930, the mooring mast at the top of the building was in place. Five more months were used to finishing up the exterior walls, windows, and other details.

On May 1, 1931, the Empire State Building was open. Smith's granddaughter cut the ribbon. President Hoover interrupted his cabinet meeting to switch on the lights of the Empire State Building from the White House. The Empire State Building was conceived during the last stretch of the prosperous 1920s but opened in the depth of the Great Depression. The occupancy of the building was so low that it was ridiculed as the Empty State Building. More than 80 years and many renovations (new elevators, air conditioning systems, etc.) later, the building is now fully occupied and stands as the symbol of American ingenuity and can-do spirit. Tourists around the world flocked to its observation decks on the 86th floor and the 102nd floor. Despite new highs reached by other high-rises in the Middle East and Far East, the Empire State Building remains the world's most famous building.

**The Sears (Willis) Tower**. When the Sears Tower was completed in 1973, it was the tallest building in Chicago, the tallest in North America, and the tallest in the world. Its roof ends at a height of 1,451 ft (442 m). Two antennas were added later that extended to a height of 1,730 ft (527 m).

The Sears Tower was planned to be the new headquarters to house the 13,000 employees of Sears Roebuck and Co. in Chicago. It was designed by the famous architect-structural engineer team of Bruce Graham and **Fazlur Khan** of the architecture firm of Skidmore, Owings and Merrill. The basic structure of the building would be a system of steel-framed tubes. Fazlur Khan was credited as the first to use a tubular design for tall buildings. Educated at University of Dacca in Bangladesh and University of Illinois, Urbana-Champaign, Kahn became a leading tall-building structural engineer by the late 1960s.

The tubular structure Graham and Khan came up with for the Sears Tower has nine square tubes, each 75 ft × 75 ft (22.9 m × 22.9 m). Each side of the tube is not a solid wall but consists of six equally spaced columns. Not all tubes go all the way to the top. Two tubes were terminated at the 50th and 66th levels, respectively, and three more were terminated at the 90th floor. Only two rise to the full height. The figure below shows the cross-section at different floors.

Because of limited street access in a busy city, the manufacturing and transportation of the steel beams and columns were carefully planned. Two horizontal members of 15 ft (4.6 m) and a vertical member of 25 ft (7.6 m) were prefabricated into a module, nicknamed "Christmas tree," and shipped to the site immediately before being welded into the building frame. This process eliminated the need for any on-site storage and quickened the pace of construction.

The Sears Tower changed its ownership in 2009 and was renamed **Willis Tower**. A new glass-bottom sky-deck was added in 2009 on the 103rd floor and became an instant hit for tourists.

In May 2013 the 104-story One World Trade Center, built at the original World Trade Center site in New York City, became the tallest building in America when its 18-piece silver spire topped out the tower at 1,776 feet, symbolizing the year The Declaration of Independence was signed.

**The Hoover Dam**. The Hoover Dam on the Colorado River is the most famous dam in the world. When it was completed in 1936, during the Great Depression, it was number one in hydropower generation in the nation with its 1,434 megawatt capacity and was also the highest dam at 221 m (725 ft). The lake the dam created, Lake Mead, is still the largest artificial lake in the nation. The Hoover Dam is a gravity-arch dam.

There are four basic types of dam construction: Embankment dam, gravity dam, buttress dam, and arch dam. Embankment dams are made of earth and rock, and sometimes are called earth dams or rock-fill dams. Its upstream–downstream direction base is larger than the height. The weight of the dam is spread over the large base. It is therefore suitable for weak foundations underneath the dam. It has a core of compacted impervious soil so that water cannot easily seep through and under it. The gravity dam, on the other hand, is made of concrete. It stands to resist the lateral water pressure by its weight and the friction at its base. It is suitable for sites with strong foundations. The buttress dam can be considered a hollowed-out gravity dam with regularly spaced buttresses replacing the solid body, thus reducing the weight.

The arch dam is a thin shell curved both in the horizontal direction and the vertical direction. It is structurally the most efficient, resisting the lateral water pressure mainly through compression in the shell body. It is suitable for narrow canyons with strong abutments. Obviously it requires the least material, mainly concrete. The design of the arch dam is the most challenging and was not perfected until the advent of digital computer.

The Hoover Dam curves horizontally like an arch dam but its cross-section is that of a gravity dam. Because it has the cross-section of a gravity dam, with a base length of 201 m (659 ft), height of 221 m (725 ft), and crest length of 379 m (1243 ft), it is massive, containing 2.48 million cubic meters (3.24 million cubic yards) of concrete. It was the most massive dam in the States when it was completed. The amount of concrete is enough to cover the entire city of Atlanta, Georgia, by over a quarter of an inch of depth. Both its power-generation capacity and its mass, however, were surpassed in 1945 by the Grand Coulee dam of Washington State. Hoover Dam was the highest dam in the States as well, until 1968 when the Oroville dam in California was completed with a height of 235 m (771 ft).

Figure 5.17 The Sears Tower (Left), Tube Structure with Simplified Floor Plans (Right).

Source: http://en.wikipedia.org/wiki/File:Sears_Tower_ss.jpg.
Copyright © Cmglee, (CC BY-SA 3.0) at http://en.wikipedia.org/wiki/File:Willis_Tower_tube_structure.svg.

Because the massive amount of concrete was poured during the construction, a special refrigeration plant was constructed to generate cold water piped throughout the dam mass to take away the heat generated by the concrete. Without such care, the excessive expansion caused by the heat and the subsequent shrinkage during the curing process would have created unacceptable cracks in the dam body.

The Hoover Dam is an engineering marvel, but its creation is the combination of engineering vision and political drama, which played out from 1922 to 1947, directly involving five presidents.

To understand the need for the Hoover Dam, one needs to understand the Colorado River first. The 2,334 km (1,450 miles) river traces its origins to the Green River in Wyoming and Lake Granby in Colorado. The Green River joins the main branch of the Colorado River in Utah and flows southwest into Arizona, then west to touch Nevada. It turns south at the Black Canyon, separates California and Arizona, enters Baja California, and eventually reaches the Sea of Cortez. Its eastern upper reach can be traced to the Little Colorado River in Arizona and further east into New Mexico. In its natural state the Colorado River flooded from time to time, causing damage in its downstream areas. From 1824 to 1904, the Colorado River flooded the Salton Basin in California at least eight times. In 1905 and 1907 it flooded the Imperial Valley in California and the Salton Basin again. The numerous floods created the Salton Sea today.

Back in 1902, a civil engineer, **Arthur Powell Davis**, had his first vision of damming the Colorado at Black Canyon. He produced a monumental engineering report and campaigned for twenty years for a dam. He became the director and chief engineer of the Reclamation Service (later renamed the Bureau of Reclamation). In 1923, out of frustration that his grand vision was bogged down in political mazes, he resigned from the Reclamation Service and went to California to work on local aqueducts, and to Turkestan, where he became the Soviets' chief consulting engineer on irrigation. When the Hoover Dam project eventually started in 1931, Mr. Davis was not on many people's mind, but in 1933, the new Secretary of Interior Harold Ickes appointed him consulting engineer for the Boulder Dam project (see below for the name change). At 72 Davis was now old and frail. He died one month after his new appointment. (See the *Fortune* magazine article of September 1933, posted on the Bureau of Reclamation website.)

The eventual naming of the dam after **Herbert Hoover** had its origin in 1922 when a commission was formed with a representative from each of the seven basin states and one from the federal government. The federal representative was the Secretary of Commerce Herbert Hoover, under President **Warren Harding**. Hoover was credited for the division of the basin states into upper basin states (Colorado, New Mexico, Utah, and Wyoming) and the lower basin states (Arizona, California, and Nevada), and worked out an arrangement with all the basin states for the distribution of the waters of the Colorado River. The resulting Colorado River Compact opened the way for the passage of the Boulder Canyon Project Act by both houses in 1928, signed into law by President Calvin Coolidge on December 21 after Herbert Hoover was elected as the new president. The Bureau of Reclamation awarded the project in 1931 to a consortium of six contractors, named simply Six Companies, Inc.. President Roosevelt dedicated the dam on September 30, 1935, two years ahead of schedule. After the dam was named Hoover Dam in 1931 by the Secretary of the Interior Ray Lyman Wilbur, under President Hoover, following past practices, Hoover lost his bid for reelection to Roosevelt in 1932. The new Secretary of Interior Harold Ickes, under President Roosevelt, renamed the dam the Boulder Dam. Not until 1947 was the name changed back to Hoover Dam by an act of Congress and signed into law by President Truman.

Today the Hoover Dam serves four major purposes: flood control, irrigation, municipal water usage, and power generation. It is also a major tourist attraction in the southwest. The Lake Mead National Recreation Area is one of the busiest among all the Park Service areas. After an upgrade in 1993, Hoover Dam's power generation capacity is now over 2,000 megawatts. The lives of tens of millions of people in the southwest of the U.S. are directly touched by the Hoover Dam every day, not to mention the many more people in other regions who enjoy the inexpensive agricultural products of the southwest made possible by the Hoover Dam. The Colorado River today disappears in the desert long before it reaches the Sea of Cortez. It is estimated 90% of its water is used for agricultural irrigation.

A lasting legacy of the Hoover Dam was the creation of a city, **Boulder City**, Nevada. It was created on the west bank of Black Canyon to house the workers for the dam project. At the peak of the construction, 5,000 workers and their family members lived in the new city. After the completion of the dam, the population dwindled significantly, but by 2013, its population is 15,000 and it became a vibrant community for recreation and leisure living.

The crest of the Hoover Dam was a part of US Route 93. After the 9/11 terrorist attacks, safety concerns prompted the design and construction of a bypass downstream. It was completed in 2010.

It is also interesting to note that the Hoover Dam is ranked 38 in the world in hydropower-generation capacity. The number one is the **Three Gorges Dam** in China at 18,300 megawatts (22,500 megawatts by 2012) and the second is the **Itaipu Dam** of Brazil at 14,000 megawatts. The Three Gorges Dam is a gravity dam across the famed Yangtze River. The dam body was completed in 2006. The Itaipu Dam, completed in 1984, is a very long dam system (7,700 m long) across the Paraná River on the border between Brazil and Paraguay. Its main dam is composed of hollow concrete segments and its wing dams are earth and rock-fill dams. In 1994 the American Society of Civil Engineering sought nominations across the globe for the Seven Wonders of the Modern World. The **Empire State Building**, the **Golden Gate Bridge**, and the **Itaipu Dam** were among the seven selected. It is interesting to note that the Empire State Building, the Golden Gate Bridge, and the Hoover Dam were all completed during the Great Depression period from 1929 to 1939.

**The Grand Coulee Dam**. The Grand Coulee Dam is located in the northeastern part of Washington State on the Columbia River. The Columbia is the second-largest river in the United States, after the Mississippi River. It originates in Canada and flows south. Approximately 100 miles (161 km) into Washington, it is joined by the Spokane River and turns westward. After another 100 miles, it turns south until it is joined by the Snake River just north of the Washington–Oregon border. From there on, it follows the border all the way to the Pacific Ocean. The section between the Spokane and the Snake rivers forms a half loop, called The Big Bend. The area east of the Big Bend is called Grand Coulee Basin; "Grand Coulee" is a geological term referring to the basin created from the erosion action of the water diverted by a glacier during the ice age. The town of Grand Coulee sits about halfway on the westward part of the Big Bend. The idea of building a dam near Grand Coulee to provide irrigation water to central Washington had been around for 30 years before Franklin D. Roosevelt's administration promised funding in 1933 and initial excavation at the dam site began in December of 1933. In 1935, Congress finally authorized full funding for the dam. The dam body was completed in 1941. In the meantime, the war effort escalated the need for electrical power in the region. More emphasis was put on power generation. By 1942 its power generation capacity was at 6,809 megawatts, the world's largest. It is still ranked fourth in the world.

The Grand Coulee dam is a concrete gravity dam with a height of 550 ft (168 m) and crest length of 5,223 ft (1,592 m). It contains 11,975,521 cubic yards (9,155,942 cubic m) of concrete, almost four times that of the Hoover Dam, and is the most massive concrete dam in the United States. Its reservoir is called Franklin D. Roosevelt Lake, which reaches to the Canadian border. Its pumping stations feed the Banks Lake south of the dam. The Grand Coulee Dam's irrigation area is about 600,000 acres (242,811 hectares).

**The Oroville Dam**. The Oroville Dam is located approximated 70 miles (113 km) north of Sacramento, California, on the Feather River. It is part of the California State Water Project and is operated by the California Department of Water Resources. Conceived in 1940 as a means to create a large water-storage

Figure 5.18 The Hoover Dam (Left), The Hoover Dam Bypass (Right).

Source: http://www.usbr.gov/lc/hooverdam/images/C45-300-021094.jpg. Copyright in the Public Domain.
Source: http://en.wikipedia.org/wiki/File:MikeO%27Callaghan%E2%80%93PatTillmanMemorialBridge.jpg. Copyright in the Public Domain.

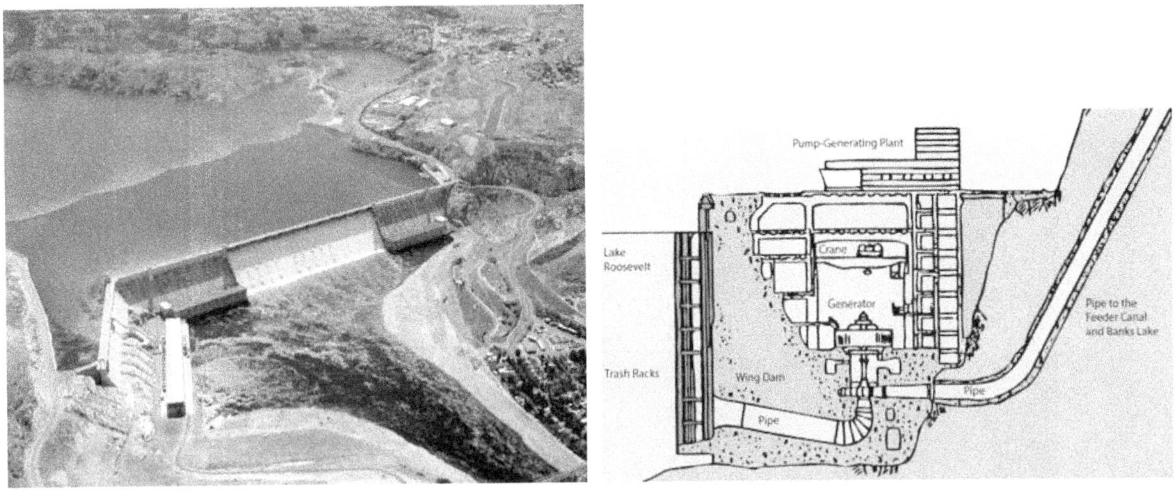


Figure 5.19 Aerial View of Grand Coulee Dam (Left), Cross-Section of the Pump-Generating Plant (Right).

Source: http://users.owt.com/chubbard/gcdam/highres/dam08.jpg. Copyright in the Public Domain.
Reclamation: Managing Water in the West, pp. 1, U.S. Department of the Interior. Copyright in the Public Domain.

reservoir to mitigate flooding of the surrounding areas, its construction did not begin until 1961, after California voters approved a bond issue in 1960 to fund the project.

The Oroville Dam is the tallest dam in the United States with a height of 770 ft (235 m) and the world's 22nd. It is an earth dam, completed in 1968. Its main function is to mitigate flooding. The Oroville Reservoir has a capacity of 3,537,577 acre-feet (436,353 hectare-m). By storing water from heavy rain or spring snow thaw and careful release of the water, it saves lives and prevents property damage. It is estimated one billion dollars of damage was averted during one major storm in 1997. Its ten power generators in two plants generate about 760 megawatts of power, mainly used to satisfy peak-hour electricity demand.

## 5.4 Modern Legends and Milestones

**The First Civil Engineer**. In the English-speaking world, the term "civil engineer" was coined by **John Smeaton** around 1770 in Britain. In the eighteenth century Britain engineers were engaging in military and non-military practices. Those specialized in military matters were called Military Engineers. By 1760 the people engaging in assessment and design of civil works projects were growing in number. By 1770 a dozen or so became well known throughout Britain. Because they were often needed to testify before Parliament on civil works, they resided in London. Among them John Smeaton (1724–1792) was the most prominent and a leader. Smeaton first identified himself as a civil engineer in 1768. In 1771, he and his colleagues founded a Society of Civil Engineers. The society remains today as a social society (renamed the Smeatonian Society of Civil Engineers after Smeaton's death). Smeaton had a good education and had talent for mechanical tools. By 1750 he started a business in improving instruments for navigation. Several papers of his on mechanical appliances were read before the Royal Society. He became a Fellow of the Society at age 29. His work then included windmills, watermills, bridges, lighthouses, and canals. The society he founded inspired the founding

Figure 5.20 The Oroville Dam.

Source: http://www.dwr.water.ca.gov/newsroom/photo/facilities-swp/oroville_dam.jpg. Copyright in the Public Domain.

of the Institute of Civil Engineers, London, 1818, which in turn served as a model for the American Society of Civil Engineers, founded in 1852.

**Safe Drinking Water.** Improving drinking water quality through some kind of treatment has been recorded for thousands of years. Ancient Greeks used filtration, boiling, etc., to reduce particles, improve taste, and reduce odor. Ancient Egyptians used the chemical alum to settle particles in water. Filtration became the main treatment during the 1700s. In 1800 Europe, slow sand filtration was used. The discovery by Dr. John Snow in 1855 that the outbreak of cholera was linked to a public well in London polluted by sewage, and the medical knowledge advanced by Dr. Louis Pasteur in 1880 that microbes caused disease, focused attention on the removal of disease-causing microbes in drinking water supplies. In 1908, chlorine was used for the first time in the Boonton reservoir, which supplied drinking water to Jersey City, New Jersey. From that point on, the low cost and effective chlorine became the main disinfectant in drinking water treatment.

Federal regulation on drinking water safety began in 1914 when the Public Health Service set standards for the microbial quality in water, focusing only on bacteria that caused contagious disease. By 1962, the

Public Health Service standards were expanded to limit 28 substances. The creation of the Environmental Protection Agency (EPA) by President Richard Nixon in 1970 marked a turning point in public awareness and government commitment to environment-related health issues. The ensuing Safe Drinking Water Act of 1974 set the most comprehensive standard for drinking water treatment.

Chlorine treatment for drinking water, however, is not without its side effects. In the mid-1970s a team of scientists and environmental engineers at the EPA conducted an exhaustive investigation and established that certain naturally occurring chemicals in the water interacting with chlorine creates disinfection by-products (DBP) that are potentially cancer-causing substances. The amount of DBPs is so small that the risk of not using chlorine to kill disease-causing germs is far greater than the risk of cancer. Nonetheless drinking water treatment plants have been trying to remove these chemicals before chlorination and/or replace chlorine with ozone, a powerful but more expensive oxidant.

**Wastewater Collection and Treatment.** In 1855 Chicago, population over 84,000, became the first major city to have a comprehensive plan for a combined domestic and street water collecting system, to rid the city of cholera and dysentery epidemics. It was credited to chief engineer **Ellis Chesbrough** (1813–1886), appointed by the newly created Chicago Board of Sewerage Commissioners. Chesbrough came with impressive credentials. A self-educated engineer, he learned surveying, grading, and tunneling while working as an apprentice on several railroad construction projects. He later worked on the construction of Boston's new water system. He was elevated in 1850 to the position of commissioner of Boston's waterworks and then the first city engineer. His achievements, including centralization of Boston's building and maintaining waterworks, sewers, streets, and harbor facilities, became well known and he was the obvious choice to solve the Chicago's wastewater problem. (See ASCE News, Illinois Section, January 2002.)

Chesbrough's plan was to discharge wastewater into the Chicago River through the web of a drainage pipe system. Because only gravity was available to drive the flow, much of the city needed to be elevated by depositing soil to create a new ground surface. Some locations were elevated as much as 16 ft to allow the gravity flow. In 1861, a new Board of Public Works was formed and Chesbrough became its first chief engineer. For twenty years, he was responsible for the city's drinking water as well as wastewater collection and treatment. He was credited for new water supply tunnels dug deep in the ground and extended far into Lake Michigan to get to unpolluted clean water. He was also responsible for the construction of the famous Chicago Water Tower, one of Chicago's landmarks.

The development of modern wastewater treatment methods follows an evolutionary path. In 1914, liquid chlorine was applied to a sewage plant for disinfection in Altoona, Pennsylvania. In 1916 San Marcos, Texas, built the first activated-sludge plant. Milwaukee, Wisconsin, built the first large-scale activated-sledge sewage treatment plant in 1919 and recycled the sewage by drying the sludge and selling it as fertilizers. In 1932 the first large-scale pilot plant for dewatering and burning filter cake was operational at the West Side plant of Chicago and all sewage sludge was dried and burned. Daytona Beach, Florida, combined coagulation and sedimentation treatment using water-plant sludge as a coagulant in 1949. In 1957 two sewage plants in Corpus Christi, Texas, used new rotary dryers for sledge disposal. In 1969 Chicago became the first city to plan to apply tertiary sewage treatment with the installation of a 15 mgd (mega-gallon-day) micro-strainer in the treatment plant at Skokie. At this point the main ingredients of a modern wastewater treatment methodology were in place. (See *Turning Points in U.S. Civil Engineering History*, Special Issue: ASCE's 125th Anniversary, ASCE, 1977.)

**The First Female Environmental Engineer. Ellen Swallow** (1842–1911) was born to a family of modest means in Dunstable, Massachusetts. Both parents were school teachers. She attended the Westport

Figure 5.21 John Smeaton (Left), Ellis Chesbrough (Middle), Ellen Swallow (Right).

Source: http://en.wikipedia.org/wiki/File:John_Smeaton.jpg. Copyright in the Public Domain.
Source: http://chicagotribute.org/Markers/Chesbrough.htm. Copyright in the Public Domain.
Source: http://en.wikipedia.org/wiki/File:Ellen_Swallow_Richards.jpg. Copyright in the Public Domain.

Academy from 1895 to 1863 and saved enough money from various domestic jobs to attend Vassar College and graduated with a B.S. degree in chemistry in 1870. She became the first woman admitted to the Massachusetts Institute of Technology that year and earned a B.S. degree in chemistry in 1873. She also earned a Master of Art degree in chemistry from Vassar College in the same year. She became a teacher in the chemistry department of MIT and continued her study in chemistry, although at the time, MIT was not ready to grant a Ph.D. to her. She married Robert Richards, head of the Department of Mining Engineering in 1875 and collaborated on numerous projects on ore analysis.

In 1884 she began working in a newly established sanitary chemistry laboratory in the Lawrence Experiment Station and conducted a large-scale water-quality study in 1887 for Massachusetts and produced the world's first water purity table. Her work led to the first water-quality standards in America, and the first modern sewage treatment plant, in Lowell, Massachusetts. In 1892 she introduced German-coined-word **Ecology** into English. In 1900 she coauthored the textbook *Air, Water, and Food from a Sanitary Standpoint*, with A. G. Woodman. In 1999 the *Engineering News Record* honored her as one of the top environmental engineering leaders in the last 125 years and named her the first female environmental engineer. (See *Engineering Legends, Great American Civil Engineers*, by Richard G. Weingardt, ASCE Press, 2005.)

**A Tunnel Named Holland. Clifford Milburn Holland** (1883–1924) was born in Somerset, Massachusetts, and attended public schools. In his teen years he proclaimed that he wanted to be a "tunnel man." He started college study in Harvard in 1902 worked part-time jobs to pay his way to a B.A. degree in 1905 and B.S. degree in civil engineering in 1906. Immediately after his graduation he worked as an assistant engineer at the Rapid Transit Commission of New York and designed and supervised construction of subways and tunnels. By 1914 he was promoted to tunnel engineer in charge of the design and construction of four subway tunnels under the East River. He was promoted to division engineer in 1916 and hired as the chief engineer in 1919 to build the **Hudson River Vehicular Tunnel**.

The need for the Hudson River Tunnel was similar to that of the Brooklyn Bridge over the East River: overcrowding of the only means of transportation between New York and New Jersey—ferries. A Joint Commission between New York and New Jersey considered a bridge crossing in 1906 but turned to favor a tunnel by 1913. Several tunnel designs were considered in the next few years but eventually the design submitted by Holland was selected and he was hired as the chief engineer for the Hudson River Vehicular Tunnel Project.

Figure 5.22 Clifford Milburn Holland.

Source: http://en.wikipedia.org/wiki/File:Cmholland.jpg. Copyright in the Public Domain.

Holland's design called for two separate 29.5 ft (9 m) diameter tubes 50 ft (15.2 m) apart: the north tube is 8,558 ft (2,608 m) long and the south tube is 8,371 ft (2,551 m) long. The most innovative part of Holland's design was the way he solved the ventilation problem. The tube section was divided into three vertical segments: the middle segment for the traffic, the upper segment for the collection of dirty air, and the lower segment for the supply of fresh air. Eighty-four giant fans housed in four ventilation towers on both sides of the river provide the blowing and suction actions for the tubes.

Construction began in 1920. As described in the April 15, 1999 issue of *Engineering News Record*, "the tunnels were driven through the Hudson riverbed by shields working in both directions that were launched from pneumatic caissons. Each shield was pushed forward by 30 hydraulic jacks that had a total force of 6,000 tons. The tubes were lined with 2.5-ft-wide cast iron rings." Holland's dedication to the project was described in the same issue: "Holland spent most of his waking hours overseeing all tunnel design and construction. But in October 1924, he suffered what was termed a nervous breakdown. Three weeks later, he was dead of a heart attack at age 41." Within two weeks the tunnel was renamed the **Holland Tunnel**. His successor, Milton Freeman died five months later, and till the completion of the project in 1927, the project was under the direction of another famed tunnel engineer Ole Singstad. (See *Engineering Legends, Great American Civil Engineers*, by Richard G. Weingardt, ASCE Press, 2005.)

**Father of Soil Mechanics**. Soil Mechanics today is either a required course or a major portion in a required course in every BSCE curriculum. It characterizes the mechanical properties of soil through rigorous theory and experiments. Much of the content of soil mechanics can trace its origin to **Karl von Terzaghi** (1883–1963). He was born in Prague when it was part of Austria. He graduated with honors from the Technical University in Graz, Austria, in 1904 with backgrounds in mechanical engineering, theoretical mechanics, geology, and highway and railway engineering.

His engineering career began in Vienna when he gradually specialized in geological problems. His work took him to Croatia and Russia and he became well known. During a short period in 1912–1913, he came to United States touring some major dam sites in the West and gathered engineering reports on the major problems encountered in the design and construction of the dams. During World War I, he became a professor in the Royal Ottoman College of Engineering in Istanbul (later renamed Istanbul Technical University) and began his research on soil properties. His first published work on retaining walls established him as the leading scholar on the mechanical behavior of soil. After the war he taught at Robert College at Istanbul and started extensive laboratory work with new instruments he invented to measure soil permeability and the interaction between soil particles and water. His theoretical development based on his measurements cumulated in a 1924 publication that earned him international acclaim. He soon accepted an offer from the Massachusetts Institute of Technology and came to the United States for the second time. At MIT he again started a new laboratory with his own instruments. His publications in 1925 and 1926 greatly enhanced the civil engineers' understanding of this new field of soil mechanics.

He returned to Vienna in 1929 and for the next ten years, he lectured and consulted widely throughout Europe, all the while continuing his research into foundation settlement and foundation improvement through grouting. In 1939 he immigrated to the United States for good. He taught at Harvard University and continued his consulting work with many major international projects, including Egypt's Aswan Dam. He died in 1963 at the age of 80. From 1930 to 1955 he was the recipient of the prestigious Norman Medal of American Society of Civil Engineers four times (1930, 1942, 1946, and 1955) "for a paper definitively contributing to engineering science." It is a record yet to be broken in the history of the Norman Medal. He was also the recipient of nine honorary doctoral degrees from eight different countries.

Another giant in the development of soil mechanics and foundation engineering is **Arthur Casagrande** (1902–1981), who happened to be also an Austrian-American. He was Terzaghi's assistant during Terzaghi's MIT years and made contributions on his own in the complex tri-axial test development and the development of other sophisticated test apparatuses. Casagrande was a superb educator. He developed numerous training programs for the Army Corps of Engineers and was responsible for the development of a successful post-graduate soil mechanics and foundation Engineering program at Harvard University. When ASCE established the Terzaghi Award "for an author of outstanding contributions to knowledge in the fields of soil mechanics, subsurface and earthwork engineering and subsurface and earthwork construction," Arthur Casagrande became its first recipient in 1963.

**The Interstate Highway System**. Officially known as the Dwight D. Eisenhower National System of Interstate and Defense Highways, its origins can be traced back to 1922, when General John Pershing, General of the Armies, produced a map entitled "Project for Development of National Highways of United States." The routes in the map were decided mainly on the need for large-scale evacuation and military significance. (See *The Interstate Highways System*, by Henry Moon, published by the Association of American Geographers, 1994.) The Federal-Aid Highway Act of 1938 called on the Bureau of Public Roads (BPR), the predecessor of the Federal Highway Administration (FHWA), to study the feasibility of a toll-financed system of three east–west and three north–south superhighways. The BPR's report concluded that a toll network would not be self-supporting and advocated a 26,700-mile interregional highway network. In 1944, a committee appointed in 1941 by President Franklin D. Roosevelt recommended a system of 33,900 miles, plus an additional 5,000 miles of auxiliary urban routes. Serious funding to carry out the recommendation would not come until 1956.

Under the leadership of President Eisenhower, two important pieces of legislation were passed and signed into law in 1956: Federal-Aid Highway Act of 1956 and the Highway Revenue Act of 1956. The first act increased the system's proposed length to 41,000 miles, required nationwide standards for design of the system, changed the name to the **National System of Interstate and Defense Highways**, and for the first time split the cost of federal/state share at 90/10 percent. The second act created the Highway Trust Fund for revenue from federal gas and other user taxes. It was the second act that guaranteed the funding of the interstate highways.

In the years after 1956, Americans' use of automobiles increased quickly and the interstate highway system has helped define the American way of life. The 2007 data of the FHWA show the total length of the system is 46,934 miles (75,533 km); the interstate highway of America is the world's longest highway system. It is remarkable that two generals, Pershing and Eisenhower, were responsible for the creation of the system and that its genesis was rooted in military/defense considerations.

**From Moment Distribution to Finite Element.** By the turn of the 20th century, the fundamental theory of structural analysis for linearly elastic beams, trusses, and frames is mostly developed. The practical application, however, was limited to statically determinant structures: structure for which static equilibrium equations alone determine the internal force distribution, for which the resulting equations can often be solved sequentially, i.e., two or three at a time. Thus hand computation was sufficient. For continuous beams and frames, however, the equations resulting from consideration of deformation often numbered in tens, even hundreds, even for relatively simple structures such as bridges and concrete buildings. For those structures only some approximate method of analysis could be applied. This situation was changed when a University of Illinois professor, **Hardy Cross** (1885–1959), published his moment distribution method in a paper entitled "Analysis of Continuous Frames by Distributing Fixed-End Moments" in 1930 in the *Proceedings of ASCE.* He later developed a similar method for the analysis of flows in pipeline networks.

The moment distribution method appeals to engineering intuition. It visualizes how a structure responds to loading. The method is an iterative method, mathematically, but it converges fast; normally after two circles the error became negligible. When the paper was formally published in the *Transactions of ASCE* in 1932, thus receiving wider notice, it became an instant hit among structural engineers because it made possible for them to solve heretofore intractable problems. Cross was awarded the Norman Medal by the American Society of Civil Engineers in 1933.

Hardy Cross was born in Virginia to parents from prominent southern families. He was an excellent student, receiving a B.A. degree in 1902 and a B.S. degree in 1903, both from Hampden Sydney College. He taught English and mathematics at Norfolk Academy before going to Massachusetts Institute of Technology to study civil engineering and earned another B.S. degree in 1908. He worked briefly as a bridge engineer and then taught at Norfolk Academy before studying at Harvard and earned a M.S. degree in 1911. Between 1911 and 1937, he taught at Brown University and then University of Illinois, and worked as a structural engineer in the Boston and New York areas. He returned to academic life at Yale University in 1937 and served as its chairman of the Civil Engineering Department until his retirement in 1950. (See "Leonard K. Eaton, Hardy Cross and the Moment Distribution Method," *Nexus Network Journal,* vol. 3, no. 3, summer 2001, <http://www.nexusjournal.com/Eaton.html>.)

In 1952 McGraw-Hill published Cross' book *Engineers and Ivory Towers,* edited by Robert C. Goodpasture. The book contains his papers and insights on engineering education and practice. His legacy is not only as the inventor of the moment distribution method but also as an insightful engineer and educator.

The advent of digital computing basically solves the problem of large numbers of linear simultaneous equations, but its application in structural analysis was limited to discrete structures such as beams, trusses, and frames. For plates, shells, and other two- or three-dimensional structures, only analytical solutions for simple geometry or finite difference approximate solutions are available. Then, in 1956, a way of discretizing an airplane wing structure was presented in a paper entitled "Stiffness and Deflection Analysis of Complex Structures," by M.J. Turner, R.W. Clough, H.C. Martin, and L.J. Topp in the *Journal of the Aeronautical Sciences.* This is one of the origins of a new method called Finite Element Method, even though the term Finite Element was not coined until 1960, when **Ray Clough** published a paper entitled "The **Finite Element Method** in Plane Stress Analysis," in the *Transactions of ASCE.* The Finite Element Method treats a continuum as an ensemble of a number of discrete elements and generates a solution, which is close to

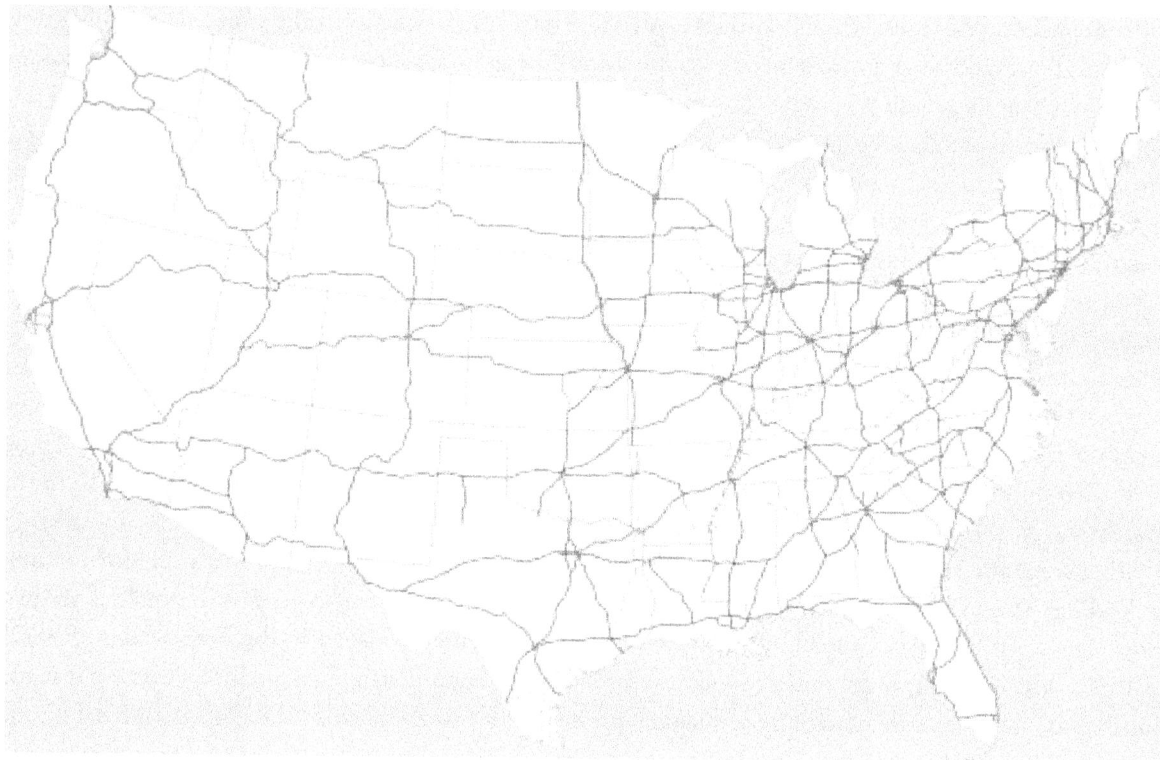

Figure 5.23 Interstate Highways in the 48 Contiguous States.

Source: https://en.wikipedia.org/wiki/File:Map_of_current_Interstates.svg. Copyright in the Public Domain.

the true solution when the elements are made smaller and smaller. Thus it is also an iterative method with a big difference: often an accurate enough solution is obtained when the elements are small enough in comparison to the general dimensions of the structure. By the late 1960s finite element analysis computer programs became available to structural engineers and the structural analysis practice was forever changed. Today, commercial programs with colorful interactive three-dimensional graphical presentations can be used to produce stress analysis results in a matter of seconds.

**A Master Engineer and Educator. Nathan Mortimer Newmark** (1910–1981) was born in Plainfield, New Jersey, and graduated from Rutgers University with special honors in civil engineering. He obtained his M.S. and Ph.D. degrees from University of Illinois in 1932 and 1934, respectively, and remained as a research assistant at the university. His time at the university as a graduate student coincided with the tenure of Hardy Cross and he studied under Cross.

During World War II he worked as a consultant to the National Defense Research Committee and served in the Pacific theater. For his excellent service he was awarded the President's Certificate of Merit in 1948. At University of Illinois he chaired the digital computer laboratory from 1947 to 1957 and helped develop the first large-scale digital computer ILLIAC II. He became the head of the Civil Engineering Department in 1956 and served for 17 years in this position. During his tenure the department's achievements and reputation reached new highs. He cared for his colleagues and young faculty. Many of them went on to become National Academy of Engineering members. His personal contribution to the field of civil engineering, especially structural engineering, were too numerous to describe. He worked in a broad range of areas and made lasting impacts. Several methods described in soil mechanics and earthquake engineering textbooks bear his name. His approach in solving a practical problem was unique in that he always zeroed in on the key factor of the problem and came up with something simple and amenable to mathematical treatment. His most-cited papers were published as a classic by ASCE in 1976 in *Selected Papers by Nathan M. Newmark: Civil Engineering Classics*, a rare honor for a living scholar.

Through his consulting engagements he left his footprint on many important civil and military projects: the Bay Area Rapid Transit (BART), Trans-Alaska Pipeline, Minute Man and MX missile systems, nuclear power plants, and most famously the Latino Americana Tower in Mexico City, Mexico. He came up with the idea of a relatively rigid structure resting on a foundation of a floating concrete box on piles to counter the poor soil conditions. The resulting 43-story high-rise (183 m/597 ft) has withstood the devastating earthquakes of 1957 and 1985 since its completion in 1956. (See *Engineering Legends, Great American Civil Engineers*, by Richard G. Weingardt, ASCE Press, 2005.)

He held many honors. He was a founding member of the National Academy of Engineering in 1964 and was elected a member of the National Academy of Sciences in 1966. He received the National Medal of Science from President Lyndon B. Johnson in 1969. ASCE established the **Nathan M. Newmark Medal** in 1975 "for his outstanding contributions in structural engineering and mechanics. The funds for the award were contributed by the honoree's former students in appreciation of the quality of education they received under his guidance at the University of Illinois." This is another testimonial to his personal achievements and his legacy as a master educator.

Figure 5.24 Nathan Newmark.

## Additional Reading Recommendations


1. *How the Great Pyramid Was Built*, Craig B. Smith, Smithsonian Books, 2004. Greater details of the building of the pyramid are given in this book.
2. *Science and Civilization of China*, Volume 4, Part 3, Joseph Needham, Civil Engineering and Nautics, Cambridge University Press, 1971. Many landmarks of China are described in this book.
3. *Engineering Legends, Great American Civil Engineers*, Richard G. Weingardt, ASCE Press, 2005. The book has a subtitle: 32 Profiles of Inspiration and Achievement.
4. *Historic Civil Engineering Landmarks of Southern California*, William Myers, The History and Heritage Committee, Los Angeles Section, ASCE, 1974.


**Image Sources:** The images in this chapter are copied from (www.Buidu.com), (www.Wikipedia.com), and (www.Google.com).

## Assignments

1. Make a three-dimensional drawing of the Khufu Pyramid using the dimensions given in this chapter. Verify the "coincidences": the square of the height of the pyramid is equal to the area of each of the inclined faces of the four triangles.
2. Write a one-page report on a modern American landmark not covered in this chapter but you believe is on a par with the ones that are, and list your sources.
3. Write a one-page report listing (a) the tallest five buildings already completed in the world, and (b) the world's longest five bridges already completed. Cite your sources.
4. The US Interstate Highway System is credited to which president? Answer: (a) FDR, (b) HST, (c) DDE, (d) JFK, or (e) LBJ.
5. It took (a) 10, (b) 12, (c) 14, (d) 16, or (e) 20 years to complete the Brooklyn Bridge.
6. It took (a) 10, (b) 12, (c) 14, (d) 16, or (e) 20 months to complete the Empire State Building.
7. It took (a) 2, (b) 3, (c) 4, (d) 5, or (e) 10 years to complete the Hoover Dam.
8. It took (a) 5, (b) 6, (c) 7, (d) 8, or (e) 10 years to complete the Holland Tunnel.
9. The Holland Tunnel was named after a (a) civil engineer, (b) president, (c) governor, (d) senator, or (e) country.
10. The first civil engineer was (a) George Washington, (b) Benjamin Franklin, (c) Thomas Jefferson, (d) Smeaton, or (e) Swallow
11. The Father of Soil Mechanics was (a) Smeaton, (b) Swallow, (c) Terzaghi, (d) Casagrande, or (e) Newmark.
12. The size of the base of the Great Pyramid is approximately equal to the size of (a) 5, (b) 7, (c) 10, (d) 12, or (e) 15 American football fields.
13. The person most responsible for the construction of the Owens River Aqueduct was (a) William Mulholand, (b) Governor Edmund G. Brown, (c) President Ronald Reagan, (d) Governor Ronald Reagan, or (e) Governor Jerry Brown.
14. A Segmental Arch bridge is called such because it (a) uses part of a semicircular shape as the main shape, (b) is constructed segment by segment, (c) has more than one segment, (d) has two arch segments, or (e) constructed by brick segments.

# Chapter 6

# Engineering Ethics

## 6.1 Overview

The word 'ethics' is defined as "the discipline dealing with what is good and bad and with moral duty and obligation," by the Merriam-Webster Dictionary. Martin and Schinzinger define ethics in their book, *Ethics in Engineering* (McGraw Hill, 4th edition, 2005), as "synonymous with morality, It refers to moral values that are permissible (all right), policies and laws that are desirable." Merriam-Webster defines 'Professional Ethics' as "the principles of conduct governing an individual or a group," while Martin and Schinzinger define 'Engineering ethics' as consists of responsibilities and rights that ought to be endorsed by those engages in engineering, and also of desirable ideals and personal commitments in engineering."

Engineering ethics are not laws in the legal sense but are laws in the sense of what are morally right and permissible in the engineering profession. From an individual conduct point of view, engineering ethics cannot be separated from individual integrity and honesty which should be learned and practiced as early as while a civil engineering student. We shall start with the student ethics issues, followed by detailed description of ethics guidelines issued by ASCE. We conclude this chapter with several actual incidents that are instructive in terms of ethics violations entailed.

## 6.2 Student Ethics

Many universities have explicit policies on 'academic honesty' and guidelines on 'student conduct.' Described herein are topics not considered serious ethics related as cheating in exams, but we recommend as good ethical practices that also benefit you.

**Attending Classes.** While the importance of attending classes was explored in Chapter 2, it was explored in terms of benefits to students. It is emphasized here that attending classes is both a privilege and a responsibility. Many students do not see the 'responsibility' side and offer comments such as "I paid the tuition to buy the education" and "I attend classes I like not ones I don't like." It must be pointed out, however, even for a private university, the tuition paid by students rarely cover the whole cost of education students received. The rest is covered by endowment funds contributed by well-wishing donors. For state universities, tuition and fees may cover some of the cost and the rest comes from tax payers' contributions. You therefore are accountable to others for your conduct.

From the university's perspective, facilities, equipments, and utilities are required for classes and instructors are hired to conduct class meetings. For instructors, going to a class requires pre-class preparation and in-class delivery and interaction with students. Skipping classes is a clear indication that the student does not appreciate the value of the class and the dedication of the instructor.

When an instructor stands in the front of a class, nothing escapes his/her notice, certainly not a couple of students keep talking to each other no matter how low their voices are. There are other behaviors that are distractive, disruptive and disrespectful not only to the instructor but to fellow students. We polled our students and they found the following disagreeable: come late and leave early, come and go at will in the middle of a class, eat and drink, use cell phone, text messaging, loudly type on laptop, cough or sneeze loudly, fart silently or loudly, yawn and fall into sleep, etc.

**Doing Homework.** We were puzzled when some students had excellent records on homework assignments but failed miserably in exams. It turned out that some homework solution manuals could be bought online and some students simply copied the solutions. Clearly coping solutions is an act of dishonesty. Even if the homework is not graded and counted toward a term grade, it is still an act of dishonesty to the student himself/herself, because the student deprives himself/herself of the opportunity to learn.

**Writing Reports.** Virtually any subject can be found on the internet and it is easy to copy in part or in whole an article already published. Citing from published work in writing a report is permitted if the report includes an element of research and the sources are identified. The word '**plagiarize**' is defined as: to steal and pass off (the ideas or words of another) as one's own; use (another's production) without crediting the source; in the Merriam-Webster Dictionary. Plagiarizing is clearly an act of unethical behavior but it is one easily caught. Just as easily to find something to copy, a simple search on the internet will find the evidence of plagiarism. Many universities offer access to faculty software that can easily identify plagiarized passages in a report.

Notice the word "ideas" in the definition above. Copying of ideas is just as bad as copying words if not worse. It is also easily caught. We identified two student reports with identical format and main contents but changed words here and there. Most universities have explicit policies dealing with student plagiarism. Some publishers have exact guidelines allowing copying up to 40 words or so without asking for permission.

**Taking Exams.** In the age of computers and internet, cheating on exams is increasingly easy. Unknown to students, however, is the fact that all cheatings are easily discovered. Severe punishments often accomplish exam cheating. Many universities have student/faculty panels to deal with exam cheating. Our experience in serving on these panels is that student members are often more strict, unforgiving and inclined to handout severe punishment than faculty. That is rightly so since a cheating incident dishonors not only the wrongdoer but the student body as a whole.

Exam cheating is not only unethical but unnecessary. Just follow the advice in Chapter 2, and you will never be in a situation that tempting cheating.

**Being Responsible in Carrying out Organizational Duties.** The importance of co-curricular activities is explored in detail in Chapter 4. Each student organization has elected officer and each elected officer carries specific responsibilities and duties. To members of the organization and to the university, these responsibilities are expected to be executed by the elected officers who are the natural leaders of the organization. Being elected an officer means one is committed to carrying out the responsibilities and duties the position carries. When officers neglect to carry out the duties, the organization ceases to function and all members suffer. We observed that the level and quality of activities of student organizations changed from one leadership group to another, sometimes drastically. It simply reflected the commitment of the leadership group. Merely enjoying being an officer but not doing what is expected is unethical. Besides, it terms of boasting one's resume, simply listing a position is not enough. People expect to know what an officer has done.

## 6.3 Guidelines to the Standards of Professional Conduct of ASCE

Engineering ethics being the ethics of a profession, each professional organization in engineering has explicit code of ethics. For civil engineers, the most pertinent is the code of ethics of ASCE. ASCE also publishes its **Standards of Professional Conduct**, the purpose of which is described as follows: "The Standards of Professional Conduct were developed to provide individuals or small businesses that don't have the resources or a complete set of principles and guidelines to govern the day-to-day aspects of ethics practices in our profession. These guidelines reinforce ASCE's Code of Ethics, which all ASCE members are expected to practice." While the code of ethics are described in more general terms, the guidelines are more specific. These **Guidelines** to the Standards of Professional Conduct are briefly described below. For details please visit http://www.asce.org/uploadedFiles/Ethics_-_New/ethics_guidelines010308v2.pdf. The language in quotation marks is originally from ASCE,

1. **Conflict of Interest.** "The best interest of the Employer or profession is at issue. You are expected to avoid any relationship, influence, or activity that impairs your ability to make objective and fair decisions when performing your jobs. When in doubt, you should share the facts of the situation with your leadership and resolve the conflict." One example is you work full time in one company but occasionally for another company on temporary basis and both companies are sending bids to win the same contract.
2. **Ensuring Legal Compliance.** "ASCE members shall conduct their actions in accordance with applicable laws and regulations. As ignorance is not a defense against violations, you need to be knowledgeable of relevant laws and regulations first and following them faithfully." One example is on burial sites preservation. Most states have burial site preservation programs. If in the course of construction, your crew digs up suspected human remains, you need to stop the work and report to authority immediately.
3. **Employees and Public Safety.** "ASCE members shall be committed to maintaining a drug and alcohol free, safe, and healthy work environment. They shall comply with applicable environmental, health, and safety laws and regulations. Most companies also have in-house regulations dealing with the same issues. For example, you are not allowed to keep alcohol in your office even you claim you never actually drink it.
4. **Workplace Quality.** "ASCE desires a workplace where its members feel respected, satisfied, and valued. Harassment, discrimination, or sexist behavior of any kind is unacceptable (in many cases it is illegal)." We learned that one supervisor was accused of workplace harassment by calling a subordinate four times a day to ask about the progress of an assignment.
5. **Use and Protection of Employer's Assets.** "Your Employer has many valued assets, such as cash, physical property, proprietary trade secrets, and confidential information. Protecting these assets against

loss, theft, and misuse is every employee's responsibility." For example you must gaining approval before taking your company issued computer home for work.

6. **Maintaining Accurate and Complete Records.** "The importance of maintaining accurate and complete records cannot be overstated." For example transactions between your company and a client should be maintained in accordance with established Employer guidelines.
7. **Gifts, Meals, Services, and Entertainment.** "It is improper for an ASCE member or family member to knowingly request, accept, or offer anything that could be construed as an attempt to influence the performance of duties or to favor a customer, supplier, or competitor that is contrary to the best interests of the Employer, its clients, or the profession." For example you are not to accept gifts from a supplier to your company's business. Some companies have explicit guidelines on the maximum monetary value of gifts that you may accept.
8. **Confidential or Proprietary Information.** "In the course of normal professional activities, ASCE members may have access to information that is proprietary, confidential, privileged, or of competitive value to the Employer. ASCE members must respect these confidences by protecting the confidentiality and security of documents and related information." For example your company's in-house design manual is proprietary and cannot be disclosed to outsiders.
9. **Outside Employment/Activities.** "Outside employment or business activities not related to the Employer must not conflict with the employee's ability to properly perform his or her work. For example, you cannot work on outside activities to the extent that you feel tired working in your company's office." This is called conflict of commitment and some companies have explicit guidelines in terms of maximum hours per week you can spend on the other commitments.
10. **Purchases of Goods and Services.** "The acquisition of goods and services from external vendors may constitute a significant portion of the Employer's annual expenditures. Adherence to established guidelines and practices governing the procurement function are critical to ensure compliance with all commercial and legal requirements and to maximize the value received from these expenditures." For example you should avoid purchasing for your company from a business owned by your family members.
11. **Bribes and Kickbacks.** "ASCE prohibits its members to offer or accept bribes, kickbacks, and other similar payoffs and benefits to or from suppliers, regulators, government officials, trade allies, or customers." It is also illegal to offer bribes in foreign countries.
12. **Relationships with Competitors.** "ASCE members should be aware that the Employer may be in a competitive environment." This means your Employer's interested must be protected when dealing with competitors.
13. **Relationships with Clients, Outside Contractors, and Consultants.** "Clients, contractors, and consultants should be treated honestly, without unfair discrimination or deception, in a manner conforming to local, state, and national laws, and consistent with good business practice." Honesty is the best policy.
14. **Environmental Protection.** "ASCE members who are aware of situations in which the Employer may not be complying with environmental laws or is improperly handling, disposing of, or otherwise discharging any toxic or hazardous substance should immediately contact the Employer." The difficult part is what if the Employer ignores you. See the next one on Whistle Blowing.
15. **Whistle Blowing.** "Whistle blowing" is when an employee reports an employer who is breaking the law … To actually whistle blow, the employee must report the illegal act outside the company to a government or law-enforcement agency … , he or she is protected by law. The employer cannot retaliate

against the employee. The employer cannot fire the employee for the whistle blowing. The employer cannot mistreat the employee for whistle blowing." The employee can still be fired for reasons other than the whistle blowing.

## 6.4 The Four Fundamental Principles of the Code of Ethics of ASCE

The Four Fundamental Principles of the Code of Ethics of ASCE are reproduced below in italic.

Engineers uphold and advance the integrity, honor, and dignity of the engineering profession by:

1. Using their knowledge and skill for the enhancement of human welfare and the environment;
2. Being honest and impartial and serving with fidelity the public, their employers and clients;
3. Striving to increase the competence and prestige of the engineering profession; and
4. Supporting the professional and technical societies of their disciplines.

## 6.5 The Seven Fundamental Canons of ASCE and Guidelines to Practice

The word 'canon' means 'an accepted principle or rule; a criterion or standard of judgment; a body of principles, rules, standards, or norms,' according to the Merriam-Webster Dictionary. The seven canons of ethics of ASCE is reproduced below. For details on ASCE's guideline to practice on each canon, see http://www.asce.org/Leadership-and-Management/Ethics/Code-of-Ethics/. The seven canons cover the following: responsibilities to the public, limits of service scope, limits on public statement, relationship with employer/client, competition with others, professional honor and integrity, and lifelong learning.

***CANON 1.*** *Engineers shall hold paramount the safety, health and welfare of the public and shall strive to comply with the principles of sustainable development in the performance of their professional duties.*
***CANON 2.*** *Engineers shall perform services only in areas of their competence.*
***CANON 3.*** *Engineers shall issue public statements only in an objective and truthful manner.*
***CANON 4.*** *Engineers shall act in professional matters for each employer or client as faithful agents or trustees, and shall avoid conflicts of interest.*
***CANON 5.*** *Engineers shall build their professional reputation on the merit of their services and shall not compete unfairly with others.*
***CANON 6.*** *Engineers shall act in such a manner as to uphold and enhance the honor, integrity, and dignity of the engineering profession and shall act with zero tolerance for bribery, fraud, and corruption.*
***CANON 7.*** *Engineers shall continue their professional development throughout their careers, and shall provide opportunities for the professional development of those engineers under their supervision.*

## 6.6 Theory and Practice of Whistle Blowing: A Case Study

While the 15th Guideline to the Standards of Professional Conduct of ASCE is clear in defining whistle blowing and the legal protection to the whistleblower, in reality it is a difficult decision on the part of the employee. In classroom discussions students often expressed the fear that an act of whistle blowing will end any good working relation with the current employer even if he/she is not fired and may also end potential employment with other employers once the employee is known as a whistleblower. These are legitimate concerns and the authors could not offer clear cut advice. In theory, however, an engineer's self interest is superseded by the interest of the employer while the public interest supersedes all other interests. If and when an illegal act by the employer is such that the employee must act to prevent any damage to the public interest, it is important for a whistleblower to act prudently to make sure he/she is on solid legal grounds. The following case study of whistle blowing may of value to those who would face similar situations (See *Ethics in Engineering*, by Mike W. Martin and Roland Schinzinger, McGraw Hill, 4th edition, 2005).

The Bay Area Rapid Transit system (BART) is a key public transportation system for San Francisco and surrounding areas. The design, construction, and operations of BART are governed by the Board of Directors of the BART District. In 1967 Westinghouse was awarded the contract to design and construct an Automatic-Train Control system (ATC) for BART. The actions of three electrical engineers of BART in 1971–72 became a classical case of engineering ethics and was detailed in the book entitled *Divided loyalties: Whistleblowing at BART,* by Robert M. Anderson, Robert Perucci, Dan E. Schendel, and Leon E. Trachtman, Science and Society: A Purdue University Series in Science, Technology, and Human Values. Trachtman Leon, editor. Vol. 4. West Lafayette (IN): Purdue University, 1980. The following much abridged description is an attempt to show only those events that are key to the whole case. The three engineers are Roger Hjortsvang, Robert Bruder, and Max Blankenzee.

1. Hjortsvang wrote an unsigned memo in November of 1971 to all levels of BART management including the general manager that summarized the problems he perceived with the ATC.
2. In January 1972, the three engineers contacted several members of the BART board of directors when their concerns were not being taken seriously by lower levels of management.
3. The three engineers consulted an external expert who wrote a report with a conclusion similar to the concerns of the three engineers. When BART tried to identify the engineers behind the unsigned memos and interviewed engineers, the three engineers denied they were the authors of the memos.
4. One of the Board of Directors, Dan Helics listened to the engineers sympathetically and took the engineer's unsigned memos and the report of the consultant and distributed them to other members of the board. Without warning the engineers he also released them to a local newspaper.
5. In February 1972, Helics convinced the engineers to appear before the whole Board of Directors but the Board was not convinced that the engineers' concerns were serious.
6. In March 1972, the three engineers were asked by BART to resign or face termination. They refuse to resign and were fired on the ground of insubordination, lying to their superiors and failing to follow organizational procedures.
7. Hjortsvang could not find full time employment for 14 months, Bruder for eight months, and Blankenzee for five months.

8. In 1974, the three engineers sued BART for damage in the amount of $875,000. The Institute of Electrical and Electronics Engineers (IEEE), the counterpart to ASCE, filed friend of the court brief supporting the three engineers' "professional duty to keep the safety of the public paramount," and citing IEEE's code of ethics that the engineers must "notify the proper authority of any observed conditions which endanger public safety and health." IEEE considered the 'public' as the 'proper authority.'
9. The attorney of the engineers advised them before the case went to trial that they could not win because they had lied to their employer. They settled out of court for $75,000.

Several things can be learned from this case. When the engineers brought the case to the Board of Directors, the case became 'Internal Whistle blowing' because they jumped the chain of command but still acted within the organization. When Helics released the documents to the local newspaper, he, not the three engineers committed an act of external whistle blowing because it became public. When the engineers denied being the authors of the memos, they lied to the employer and saw the seed for weakening their legal case later. While the whistleblowers suffered economically and mentally they were able to gain employment eventually. The BART did improve the ATC system and maintained a good safety record since 1972.

## 6.7 The Collapse of the Kansas City Hyatt Regency Walkway

**The Structure.** The Kansas City Hyatt Regency Hotel was opened in 1980. The hotel has a 40-story tower, a function block and a large open atrium approximately 117 ft (36 m) by 145 ft (44 m) in plan and 50 ft (15 m) high. There were three suspended walkways connecting the tower and the function block: The second and fourth floor walkways on the west side and the third floor walkway on the east side.

The report is based on the original investigative paper published in the July 1982 issue of *Civil Engineering* magazine by the same author. The following description is an abbreviated account of what was described in this report, the *Civil Engineering* paper and other open sources.

**The Event.** On July 17, 1981 while there was a tea dance in progress in the hotel atrium, two suspended walkways, from the second and fourth floor on the west side of the atrium of the hotel collapsed. The collapse resulted in the second floor walkway falling to the ground and the fourth floor walkway piled up on top of the second floor walkway. The third floor walkway on the opposite side of the atrium was not involved in the accident. The total casualty: 114 dead and 185 injured most from the people on second floor and the ground floor under the walkway.

**The Investigation.** In the aftermath of the incident, the Mayor of Kansas City requested an independent investigation by the National Bureau of Standards (NBS, known as National Institute of Standards and Technology since 1988) for the cause of the collapse.

The NBS team did on site inspection, gathered walkway debris, all relevant documents relating to the design and construction of the hotel (in service about one year), and photographs, videotapes and records available from the media. NBS also did extensive laboratory tests on material strength from the debris and similar material and conducted analytical studies on the as-built capacity of the walkways and the estimated load at the time of the incident. Fortunate for the NBS team a TV crew was videotaping the tea dance and

Figure 6.1 The Suspended Walkways Disappeared (Left) and the Disaster (Right).

Source: http://en.wikipedia.org/wiki/File:Hyatt_Regency_collapse_end_view.PNG. Copyright in the Public Domain.
Source: http://en.wikipedia.org/wiki/File:Hyatt_Regency_collapse_floor_view.PNG. Copyright in the Public Domain.

the tape included a segment on the second floor walkway and the people on it just ten minutes before the incident. This piece of tape gave the NBS team a reliable means to estimate the number of people on the two walkways and in turn a realistic estimate of the live load at the time of the collapse. In the course of the investigation it was discovered that the original design of the suspension rod and box-beam connection by the engineers of the firm Jack D. Gillum and Associates was altered during the construction by the contractor Havens Steel. We draw the as-built configuration and the original design configuration of the connection at the 4th floor walkway below, Figure 6.2, based on the schematics and descriptions of the report.

The local details of the walkway connector are shown in Figure 6.3.

This change in the hanger rod design essentially doubled the load on the nut connecting the top hanger rod at the 4th floor to the floor beam as it carried not only the weight on the 4th floor walkway coming from the floor beam but also that on the second floor walkway coming from the hanger rod below. A detailed analysis by the NBS team revealed other problems as well.

**The Findings.** The NBS team's findings were published in the July 1982 issue of the Civil Engineering magazine by Dr. Edward Pfrang. Only the major conclusions are described below.

1. The estimated load at the 4th floor connections exceeded the capacity of the connections at all the connection locations. Any of the connection could have initiated the collapse.
2. The as-built connection and the fourth floor hanger rod did not satisfy the design provisions of the Kansas City Building Code.
3. If the original design were used, the connection would have the capacity to resist the estimated load at the time of the collapse.

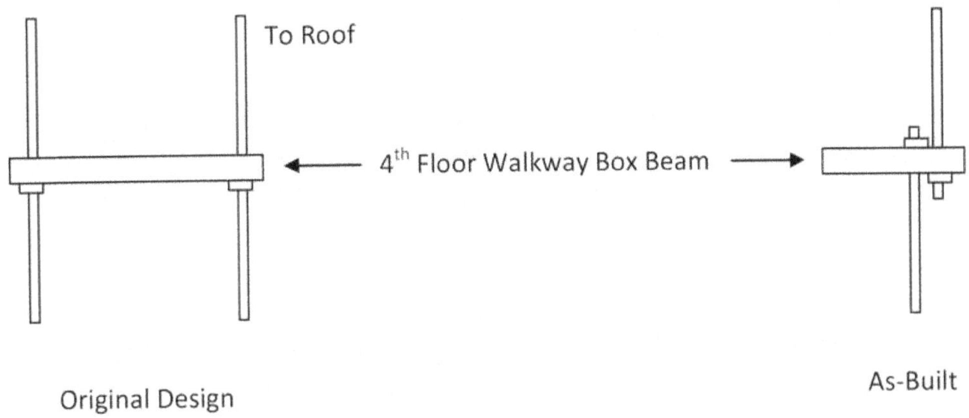


Figure 6.2 Schematics of the Walkway.

4. The connection capacity in the original design was only 60% of that of required in the Kansas City Building Code.

In other words, even the original design was flawed although it would have resisted the load at the time of the collapse. Furthermore, the design change was approved by the engineers at the firm of Jack D. Gillum and Associates apparently without a thorough review.

**Aftermath.** The engineers at the firm who approved the final drawings eventually were convicted by the Missouri Board of Architects, Professional Engineers, and Land Surveyors of 'gross negligence, misconduct, and unprofessional conduct in the practice of engineering' and their professional engineer licenses were revoked. The firm Jack D. Gillum and Associates was cleared of criminal negligence but lost its license as an engineering firm.

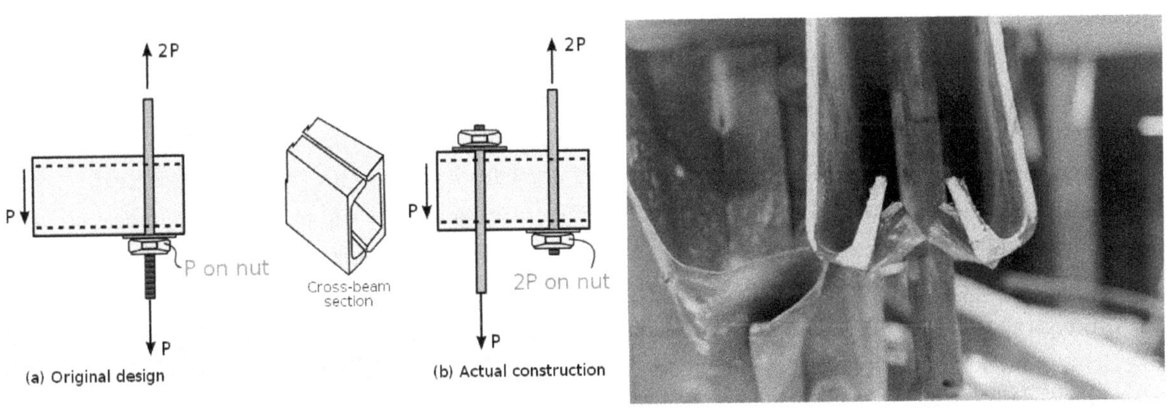


Figure 6.3 Details of the As Built and Original Hanger Rod Arrangement of the Walkway (Left) and the Damage (Right).

Source: http://en.wikipedia.org/wiki/File:HRWalkway.svg. Copyright in the Public Domain.
Source: http://en.wikipedia.org/wiki/File:Hyatt_Regency_collapse_support.PNG. Copyright in the Public Domain.

## 6.8 The Collapse of the I-35W Bridge in Minnesota

**The Structure.** The I-35W Bridge in Minneapolis, Minnesota is a fourteen-span 1,907 feet (580 m) long eight-lane bridge spanning the Mississippi River, opened in 1967. The bridge runs approximately south-north at the river crossing. The three main spans are of the deck truss type, 1,064 feet (304.3 meter) long. The other eleven spans are steel girder type. The bridge was designed by the engineering consulting firm of Sverdrup & Parcel and Associates, Inc., of St. Louis, Missouri, which was acquired in 1999 by Jacobs Engineering Group, Inc. The bridge design was certified by the Sverdrup & Parcel project manager on March 4, 1965 and approved by the Minnesota Department of Transportation on June 18, 1965. The design was based on the 1961 American Association of State Highway Officials (AASHO) *Standard Specifications for Highway Bridges* and 1961 and 1962 *Interim Specifications*, and on the *1964* Minnesota Highway Department *Standard Specifications for Highway Construction*. By August 1, 2007 an ongoing repair and renovation project closed four of the eight travel lanes (the two outside lanes northbound and the two inside lanes southbound) to traffic.

**The Event.** On 6:05 p.m., August 1, 2007 while the traffic was bumper-to-bumper on the bridge, the main spans suddenly collapsed and fell to the river and the south and north river banks. A total of 111 vehicles were on the collapsed spans. The collapse's casualty: 13 people died, and 145 people injured. The following photos are copied from the Accident Report Collapse of the I-35W Highway Bridge, Minneapolis, Minnesota, August 1, 2007, National Transportation Safety Board, NTSB/HAR-08/03, PB2008-916203. (http://www.dot.state.mn.us/i35wbridge/ntsb/finalreport.pdf)

**The Investigation.** Representatives of the National Transportation Safety Board (NTSB) and the Federal Highway Administration (FHWA) performed extensive investigative work themselves but made the consulting firm Wiss, Janney, Elstner Associates Inc. (WJE) part of the investigative team. WJE is known for its expertise in construction technology and failure evaluation. According to Howard Hill, director of technical operations and a principal of WJE (see the 12/30/2008 issue of the Engineering News Record), on site investigation by him and NTSB and FHWA experts quickly led to the gusset plates at the truss node U10, mainly because "1) that the 3-span structure separated rather cleanly along two symmetric lines, which coincided with the U10 nodes; 2) that the U10 gusset plates at all four locations … came apart in similar ways and 3) that the positions of the U10 node elements were such that it would have been difficult for them to have sustained damaged as a result of the collapse."

The numbering system for the nodes uses U for upper chord and L for lower chord and designates the southernmost node the node number zero and progresses by one going north. In the meantime, because of symmetry, the mirror image numbering is used for the northern half but with a prime (') attached.

Because the gusset plate connects the top chord members to each other and to the web members on both sides of the members, there are two gussets plates at the west side of the bridge deck and two at the east side, with a total of four at node U10.

Although the U10 gusset plates were suspect at the beginning of the investigation, the NTSB team did an exhaustive study on all aspects and possibilities. It reviewed the three renovations of the bridge and the effects on the load: The 1977 Renovation increased deck thickness resulting in an increase of dead load by 3 million pounds or 13.4 percent; the 1998 renovation resulting in an increase by about 1.13 million pounds, or 6.1 percent; the ongoing 2007 renovation brought on deck some equipment but some 250,000 pound of concrete was already milled away from the center span deck. The NTSB team examined documents from

Figure 6.4 The collapsed main span of the I-35 W Bridge.

Figure 14, Highway Accident Report: Collapse of I-35W Highway Bridge, Minneapolis, Minnesota, August 1, 2007, National Transportation Safety Board. Copyright in the Public Domain.

a 1999 inspection and discovered photos showing 'bowing' of the gusset plates at node U10, a distortion of the gusset plates (See Figure 6.6).

The team performed finite element analysis of the truss structure as well as the detailed deformation of the gusset plates. The team performed material testing on the truss members and gusset plates and confirmed that the gusset plates used were of 50,000 pound per square inch (psi) capacity. The team's study all ruled out other factors such as corrosion damage found on the gusset plates at the L11 nodes and elsewhere, fracture of a floor truss, preexisting cracking in the bridge deck truss or approach spans, temperature effects,

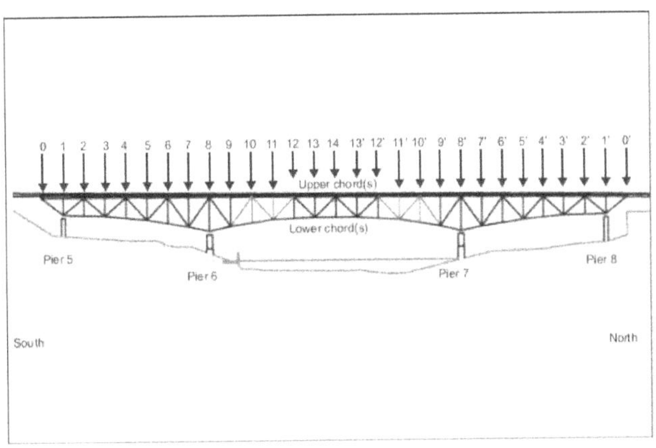


Figure 6.5 The Numbering System of the Members of the I-35W Bridge.

Figure 8, Highway Accident Report: Collapse of I-35W Highway Bridge, Minneapolis, Minnesota, August 1, 2007, National Transportation Safety Board. Copyright in the Public Domain.

Figure 6.6 Two Views of the U10 Gusset Plates.

Detail from: Figure 19, Highway Accident Report: Collapse of I-35W Highway Bridge, Minneapolis, Minnesota, August 1, 2007, National Transportation Safety Board.

or shifting of the piers, as causes for the collapse. The investigation was concluded and NTSB published a report on November 14, 2008, entitled 'Collapse of I-35W Highway Bridge, Minneapolis, Minnesota, August 1, 2007.'

Some of the major findings in addition to what have already been mentioned above are selectively reproduced below (see http://www.dot.state.mn.us/i35wbridge/ntsb/finalreport.pdf).

1. The initiating event in the collapse of the I-35W bridge was a lateral shifting Instability of the upper end of the L9/U10W diagonal member and the subsequent failure of the U10 node gusset plates on the center portion of the deck truss.
2. The gusset plates at the U10 nodes, where the collapse initiated, had inadequate capacity for the expected loads on the structure, even in the original as-designed condition.

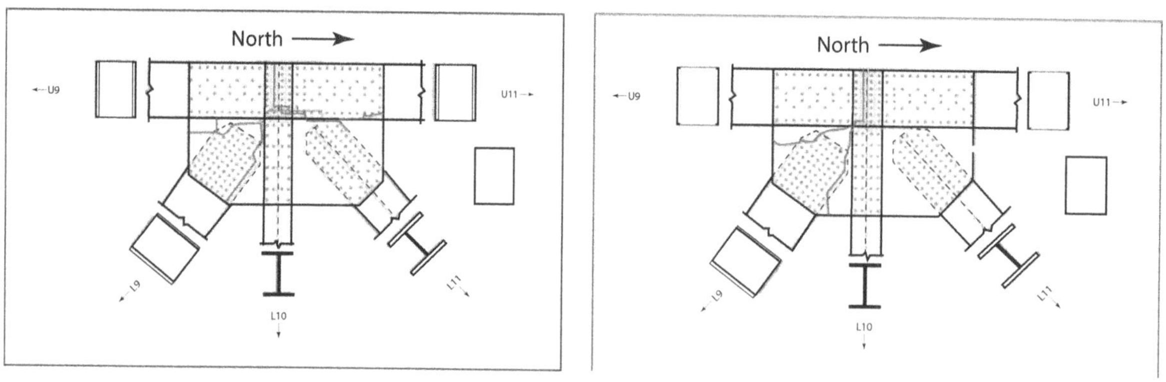


Figure 6.7 The Fracture Patterns of the U10 Gusset Plates.

Figure 20A, Highway Accident Report: Collapse of I-35W Highway Bridge, Minneapolis, Minnesota, August 1, 2007, National Transportation Safety Board. Copyright in the Public Domain.

3. Because the bridge's main truss gusset plates had been fabricated and installed as the designers specified, the inadequate capacity of the U10 node gusset plates had to have been the result of an error on the part of the bridge design firm.
4. Even though the bridge design firm knew how to correctly calculate the effects of stress in gusset plates, it failed to perform all necessary calculations for the main truss gusset plates of the I-35W bridge, resulting in some of the gusset plates having inadequate capacity, most significantly at the U4 and U4', U10 and U10', and L11 and L11' nodes.
5. The design review process used by the bridge design firm was inadequate in that it did not detect and correct the error in design of the gusset plates at the U4 and U4', U10 and U10', and L11 and L11' nodes before the plans were made final.
6. Neither Federal nor State authorities evaluated the design of the gusset plates for the I-35W bridge in sufficient detail during the design and acceptance process to detect the design errors in the plates, nor was it standard practice for them to do so.
7. Current Federal and State design review procedures are inadequate to detect design errors in bridges.
8. The loading conditions that caused the failure of the improperly designed gusset plates at the U10 nodes included substantial increases in the dead load from bridge modifications and, on the day of the accident, the traffic load and the concentrated loads from the construction materials and equipment; if the gusset plates had been designed in accordance with American Association of State Highway Officials specifications, they would have been able to safely sustain these loads, and the accident would not have occurred.

The report gave the following 'probable cause':

> "The National Transportation Safety Board determines that the probable cause of the collapse of the I-35W bridge in Minneapolis, Minnesota, was the inadequate load capacity, due to a design error by Sverdrup & Parcel and Associates, Inc., of the gusset plates at the U10 nodes, which failed under a combination of (1) substantial increases in the weight of the bridge, which resulted from previous bridge modifications, and (2) the traffic and concentrated construction loads on the bridge on the day of the collapse. Contributing to the design error was the failure of Sverdrup & Parcel's quality control procedures to ensure that the appropriate main truss gusset plate calculations were performed for the I-35W bridge and the inadequate design review by Federal and State transportation officials. Contributing to the accident was the generally accepted practice among Federal and State transportation officials of giving inadequate attention to gusset plates during inspections for conditions of distortion, such as bowing, and of excluding gusset plates in load rating analyses."

The report also made recommendations on a bridge design quality assurance/quality control program to approve final design, modifications on bridge inspection training, and including gusset plate as a structural element for design review and inspection.

A new replacement bridge at the same location, named The I-35W Saint Anthony Falls Bridge, was opened on September 18, 2008, three months ahead of schedule. The bridge is wider with ten lanes and is a box girder bridge constructed with high strength concrete.

## Additional Reading Recommendations

1. *Engineering Ethics, Concepts & Cases,* Charles E. Harris, Jr., Michael S. Pritchard, and Michael J. Rabins, 3rd Ed., Thomson Wadsworth, 2005. The book contains 70 cases in 16 categories.
2. *Engineering Ethics,* Charles B. Fleddermann, 3rd Ed., 2008. Pearson Prentice Hall, The book contains an appendix with brief description of codes of ethics of four professional Societies, including that of ASCE.

## Assignments

1. You as a civil engineer must consider the interest of (a) the public, (b) civil engineering profession, (c) your client, (d) your employer, or (e) yourself as the highest priority.
2. Engineering ethics rules depend on (a) culture, (b) place, (c) era, (d) culture, place, and era (e) none of culture, place, and era.
3. You are the onsite supervisor of a construction crew working in a city and are informed that the bulldozer dug up human-like bone remains. What should you do? Answer: (a) halt the digging and contact the city authority to report, (b) ignore it because you are behind schedule and the bones may not be human, (c) ask one of your friends who works in the city morgue to come to take a look, (d) ask one of the older crew members what to do, (e) halt the digging for three days to see if anyone is complaining, resume the work if no one is.
4. You have submitted a bid for a construction project. The owner informs you that he received several close bids and if you lower your bid price by 5%, you will be the winner and he advised you to submit a new bid. What should you do? Answer: (a) do as he suggested, (b) do as he suggested but use a lower cost material to cover the 5% difference, (c) withdraw your bid and do not want to be involved anymore, (d) do not change your bid, tell him you wish to be picked on your quality and not just price, (e) complain to the owner that he should not try to force you to cut price.
5. One of your friends is an electrical engineer who works freelance and is now considering bidding for a project that requires a professional engineer's license. You are a licensed civil engineer but he does not have a license. He asks you to help him by stamping (using your license) for him. What should you do: (a) tell him to go ahead, (b) tell him you cannot help, (c) tell him you got an A in Electric Circuits back in college and you should be the one to take on the project, (d) tell him you can help only if he pays you handsomely, (e) tell him you can help but he should not tell anybody that you are a civil engineer not an electrical engineer.
6. During a class meeting of a general education course you suddenly have an urge to tell a funny short story you heard last night to your friend sitting next to you. You should (a) speak to her in a low voice so that you feel no one is disturbed, (b) send her a text message, (c) send her an email and wait her to discover the funny email, (d) send her an email and signal her to check her email, (e) suppress the urge and concentrate on the lecture.
7. You have an early morning class at 9 a.m. that leaves you no time for breakfast. You should (a) bring a snack to class to eat, (b) bring a snack to class but sit in the last row to eat quietly, (c) ask the instructor

for permission and then bring a snack to eat, (d) go hungry through the class, (e) get up earlier to make time for breakfast before the class.

8. One of ASCE's 15 Guidelines to the Standards of Professional Conduct is entitled Workplace Quality. It addresses (a) indoor air quality, (b) indoor noise level, (c) minimum office space, (d) unacceptable discrimination, (e) working fire alarm system.
9. The guideline for the 7th Fundamental Canon of ASCE states: Engineers shall continue their professional development throughout their careers, and shall provide opportunities for the professional development of those engineers under their supervision. This refers to the importance of (a) getting a professional engineer's license, (b) being kind to engineers junior to you, (c) lifelong learning, (d) getting a Ph.D. degree, or (e) teaching part-time while working full time.
10. Offering bribes to get a contract is illegal for a US company in (a) USA only, (b) outside of USA, (c) USA and foreign countries, (d) a hostile country, (e) a friendly country.

# Chapter 7

# Life Beyond the BSCE Degree

## 7.1 Overview

Receiving a BSCE degree is a major milestone in your career. It by no means signals the end of learning. To the contrary, it simply signals the beginning of a pathway to a rewarding career, which in all likelihood requires lifelong learning to keep yourself up to date on new development and expand the portfolio of your skill set. In other words you are expected to continue your professional development throughout your career. Most new BSCE degree recipients are eager to get into the real world and gainful employment. The opportunities in civil engineering for a BSCE degree holder are described first in this chapter. Other career paths and higher degree programs are to follow. But first, one must know how to look for a job that requires a BSCE degree.

## 7.2 The First Job

In seeking employment, one question comes up often: should one looking for work in the private sector or in the public sector. That is a question difficult to answer. In terms of salary, generally private companies pay higher but some government agencies pay very well too and with better benefit and job security. Initial salary should not be a major concern although one should always know the salary range for employee of the organization and potential for promotion. For a new BSCE degree holder it is important to consider future career development opportunities beyond the first job. The questions to ask are whether there are potential exposures to a variety of different projects so that a new engineer can gain extensive experience from the work and what the career advancement opportunities within the organization are.

Whether one can relocate is a major factor in potential job opportunities. Obviously willingness to relocate broadens the prospect of choices. Ultimately, it is one's personal preference and the availability of opportunities that determine the first job.

No matter what is one's preference, job opportunities do not just appear. One must look for them. There are three types of resources one can explore while still in school:

1. **Career office.** Most universities have a career office or placement office serving graduating seniors and alumni. Many employers contact career offices for job posting or to participate in university's job fair during which many employers set up stations to highlight their job needs and accept applications on site. Staffs in career offices also give seminars on resume preparation and job interview techniques. Some even offer to review one's resume and conduct mock interviews. Most career offices also subscribe to online job search networks which maybe more useful than some free online resources. The career office is a great resource for students and one should get in touch with the office as early as possible, but certainly no later than six months before graduation.
2. **Professors.** In job hunting nothing is better than an inside track. Many professors practice engineering consulting and have extensive contact with potential employers. They may know job opportunities even before the opening is made public. Their recommendations carry a lot of weight. The relationship with one or more professors is not something to be developed just before one begins to think of jobs. In Chapter 2 the importance of professors has already been pointed out. Good grades and good interaction in and out of classroom is the key to being on the good side of a professor. Some BSCE programs hire practicing engineers as part-time instructors. They may be more important as far as job recommendations go. Certainly they have firsthand knowledge of the job market and specific job openings.

   After getting the first job, the most important thing is to keep and do well on the first job. While most civil engineers may change jobs several times in their whole career, a soured experience on the first job is damaging to the prospect of a rewarding career. In addition to the usual professionalism expected of civil engineers: punctual, honest, diligent, willing to work hard or over time, etc., one should observe the culture of the workplace: how people dress, how people interact, and how team work functions. It is advisable to have a mentor, a more senior colleague, who is willing to offer advices. One can ask the direct supervisor to recommend a mentor/advisor. A new employee is not expected to be a leader immediately in any team, but that does not prohibit you from taking initiative and request additional work assignment to assert your abilities. If and when one decides to leave the first job for career advancement, it is important to leave on friendly terms with all colleagues. They may be one's best referees for future career advancement opportunities.
3. **Personal Network.** Virtually all civil engineering or engineering related student organizations on campus have industrial advisors. If one has been an active member of a student organization or even has served on one or more officer positions, one must have already developed a relationship with the industrial advisors. Such a relationship can also be developed with practicing engineers when they are lecturing on campus in seminars or student organization activities. The more practicing engineers one knows and has a good relationship with the better opportunities one has when it comes to job hunting.

## 7.3 Some Major Companies

There are literally tens of thousands of engineering companies that hire BSCE degree holders. Briefly described below in alphabetic order are just a few who have broad ranges of engineering services and offices worldwide.

**AECOM.** The company describes itself as "a global provider of professional technical and management support services to a broad range of markets, including transportation, facilities, environmental and energy. With more than 43,000 employees around the world, AECOM is a leader in all of the key markets that it serves. AECOM provides a blend of global reach, local knowledge, innovation and technical excellence in delivering solutions that enhance and sustain the world's built, natural and social environments. A *Fortune 500* company, AECOM serves clients in more than 100 countries and had revenue of $5.9 billion during the 12-month period ended March 31, 2009." It has offices worldwide and in 42 states and D.C., Puerto Rico and Guam in the United States. It is headquartered in Los Angeles, CA with 56 other locations in California alone.

**Black & Veatch.** The professional services offered by Black & Veatch include asset, management, climate change solutions, construction, design—build/ engineering, procurement and construction, engineering consulting, engineering & design, infrastructure planning, management consulting, procurement, program management, smart utility, and water and wastewater refurbishment. Its market fields include energy, water, telecommunications, management consulting, federal, and environmental. Its 9,600 professionals work in 100 offices worldwide. In the United States, it has 65 offices in 29 states. It is headquartered in Overland Park, Kansas.

**Bechtel.** Bechtel offers services in Construction, Development & Financing, Engineering & Technology, Procurement, Project Management, Safety, and Sustainability & Environment. Its wide range projects include works on rail, road, bridges, tunnels, airport, ports, wireless and other communication infrastructure, defense, space, energy management, environmental restoration and remediation, mining and metals, oil, gas, petrochemical, liquefied natural gas, pipelines, industrial facilities, and fossil and nuclear power plants. Headquartered in San Francisco, California, it has nine other offices in Unites States and 25 offices in 24 other countries.

**CH2M Hill.** Known initially for its leadership in environmental engineering, CH2M Hill's services now cover markets in chemicals, manufacturing, electronics and advanced technology, mining, energy, nuclear, enterprise management solution, power, environmental, transportation, government and commercial facilities, water, wastewater and water resources and life sciences.

It has 25,000 employees in 40 countries. In the United States it has more than 100 project and area offices.

**FLUOR.** Founded as a construction company in 1912, FLOUR's services now include engineering, procurement, construction, maintenance (EPCM), and project management. FLOUR is a *Fortune 500* company with global reach. The range of markets FLOUR services includes energy and chemicals, industrial and infrastructure, government, power, and global services. Headquartered in Irving Texas, it has 41 offices outside of the Unite States and 22 offices in 12 states and Puerto Rico in the United States.

**JACOBS.** A broad-based technical professional consulting firm, JACOBS offers services in architecture/engineering, construction, **environmental, health and safety, planning, management, m**odular fabrication, and technology. Its market covers aerospace and defense, automotive and industrial, buildings, chemicals and polymers, consumer and forest products, environmental programs, infrastructure, oil and gas, pharmaceuticals and biotechnology, refining, and technology. It has 123 offices in 37 states and D.C. in the United States and 160 offices in over 26 countries.

**KBR.** A technology-driven engineering, procurement and construction (EPC) company, KBR is headquartered in Houston, Texas, the energy capital of the world. It supports the energy, petrochemicals, government services and civil infrastructure markets. KBR has the following business units: downstream, government and infrastructure, services, technology, upstream, and ventures. It has offices in Asia, Europe, Australia, North America and South Africa.

**PARSONS.** An engineering and construction company, PARSONS offers services in asset management, commissioning, qualification, validation, condition assessment, construction, data management, design, development and fabrication, disaster response, intelligent/security. Operations and maintenance, planning, project management and construction management. Its market covers communications, education, energy, environment, facilities, federal government, health care, infrastructure, transportation, vehicle inspection, and water and wastewater. Its projects in 34 states in United States and 15 other countries worldwide.

**PARSONS BRINCKERHOFF.** A firm in the development and operations of infrastructure, Parsons Brinckerhoff provides services in strategic consulting, planning, engineering, and program and construction management in multiple market sectors, including transportation, power, buildings/facilities, water/wastewater, environmental, and urban/community development. It has offices in Australia, Asia, Middle East, Africa, Europe, South America and North America. In the United States, it has offices in 64 cities, with its headquarters in New York City, New York.

**URS.** A global, fully integrated engineering, construction and technical services firm, URS offer s professional planning, design, environmental, construction, program and construction management, operations and maintenance, management and specialized technical services. Its market includes transportation, power, industrial infrastructure and process, environmental and nuclear management, facilities, water/wastewater, mining, and defense and security programs. It has more than 300 offices and job sutes in major cities in 31 countries in the Americas, Asia-Pacific, the Middle East and Europe. URS has offices in every state of the United States.

The above are a sample of companies with wide range of services in civil engineering and other fields. There are many other companies specialize in one or two technical areas in civil engineering serving global, regional or local markets. These can be searched on internet or the database in career offices.

## 7.4 Some Major Federal and State Agencies

Virtually every federal or state agency hires civil engineers but some agencies hire more because their responsibilities are closely related to civil engineering. A sample of four of these agencies is given below. Their responsibilities and operations cover most of the civil engineering technical areas.

**Army Corps of Engineers.** The United States Army Corps of Engineers (USACE) is one of the oldest federal agencies established by the Congress. Its mission is to "provide vital public engineering services in peace and war to strengthen our Nation's security, energize the economy, and reduce risks from disasters." Today it "is the Nation's number one federal provider of outdoor recreation, owns and operates more than 600 dams, operates and maintains 12,000 miles of commercial inland navigation channels, dredges more than 200 million cubic yards of construction and maintenance dredge material annually, maintains

926 coastal, Great Lakes and inland harbors, restores, creates, enhances or preserves tens of thousands of acres of wetlands annually under the Corps' Regulatory Program, provides a total water supply storage capacity of 329.2 million acre-feet in major Corps lakes, owns and operates 24 percent of the U.S. hydropower capacity or 3 percent of the total U.S. electric capacity, supports Army and Air Force installations, provides technical and construction support to more than 100 countries, manages an Army military construction"

USACE serves the public and the military but it also serves private firms by providing engineering and technical support to U.S. firms competing for contracts for overseas projects on cost reimbursement basis. It commands approximately 37000 civilian and military employees working in over 70 countries worldwide. In the United States its offices spread geographically in ten Divisions and 37 Districts.

**Environmental Protection Agency.** The mission of the United States Environmental Protection Agency (USEPA) is "to protect human health and the environment," and to "lead the nation's environmental science, research, education and assessment efforts." Although it is a regulatory agency, its science and technology laboratories and research centers hire civil and environmental engineers. Its more than twenty laboratories and centers are located in 16 states.

Each state has its own Environmental Protection Agency charged with developing, implementing and enforcing the state's environmental protection laws. They also hire civil and environmental engineers.

**Department of Transportation.** The mission of the United State Department of Transportation (USDOT) "is to serve the United States by ensuring a fast, safe, efficient, accessible and convenient transportation system that meets our vital national interests and enhances the quality of life of the American people, today and into the future." Of its 12 federal agencies at least the following eight are closely related to civil engineering and hires BSCE degree holders: Federal Aviation Administration (FAA), Federal Highway Administration (FHWA), Federal Railway Administration (FRA), Federal Transit Administration (FTA), Maritime Administration (MARAD), National Highway Traffic Safety Administration (NHTSA), Research and Innovation Technology Administration (RITA), and Saint Lawrence Seaway Development Corporation (SLSDC).

Many more job and career opportunities are in each state's department of transportation (DOT) or equivalent (e.g. Caltrans for California). Most state DOTs carry out design, construction and maintenance themselves although more and more they also subcontract out their work. In Chapter 5 the funding for the interstate highway system was described. The funding is allocated to the state level annually, making the state DOTs one of the most stable and reliable employers for BSCE degree holders.

**Bureau of Reclamation.** The US Department of Interior's Bureau of Reclamation (USBR) is charged with managing the water resources in the west part of United States. The 16 western states are covered by five regions under USBR: Great Plains, Lower Colorado, Upper Colorado, Mid-Pacific and Pacific Northwest. A total of 30 offices and the D.C. headquarters house administrators and engineers. The USBR's activities cover building seismic safety, canal safety, dam safety, desalination, fisheries and wildlife resources, flood hydrology, geotechnical engineering, hydroelectric research and technical services, infrastructure services, land and water surfaces use, materials engineering and research, national irrigation water quality, recreation, remote sensing and GIS, river systems and meteorology, sedimentation and river hydraulics, stream corridor restoration, water conservation, water operations, water resources research, and water resource services.

Many cities and counties, especially those in major population centers, have engineering offices with a large professional staff. The water and wastewater treatment is the responsibility of local governments, so is solid waste collection and disposal. While some local governments contract out these services to private companies, many also create semi-public entities to manage water supply and wastewater treatment. These entities hire many civil engineers.

## 7.5 PE Exam

Beyond the BSCE degree the next personal milestone for a civil engineer is the licensing as a professional engineer. The licensing of professional engineer is regulated by each state. The last hurdle for a civil engineer to obtain the professional engineer license is to pass the Principle and Practice of Engineering (PE) exam. It generally cannot be taken right after the BSCE degree. In fact application to take the exam requires the passing of the FE exam described in Chapter 3, a BSCE degree (or any engineering degree from an ABET accredited program), and four years of practice under the supervision of professional engineers. The four years of practice after the BSCE degree is considered as internship for an engineer-in-training period. For Master's degree holders most state cuts the 4-year requirement to two. A study of the passing rate of the PE exam from between October 2005 and October 2008 by the National Council of Examiners for Engineering and Surveying (NCEES) reveals that the passing rate is the highest for exam takers with four and five years of experience (70 and 68%, respectively). This was reported in the June 2009 NCEES newsletter *Exchange* in a leading article entitled 'PE pass rate demonstrate importance of experience.'

While the PE exam can be taken by any qualified engineering degree holders, the content of the exam is discipline specific. For civil engineers, the one to take is the PE Civil exam. The PE exam is similar to the FE exam only in the length of the exam; four hours of morning session and four hours of afternoon session separated by a one hour lunch break. The morning session of the PE Civil exam emphasizes the breadth and includes the follow five basic areas of civil engineering: Construction, Geotechnical, Structural, Transportation, and Water Resources and Environmental. The afternoon session emphasizes depth. Exam takers are to choose only one of the above five areas. The breadth exam and the depth exam contain 40 multiple-choice questions each. Only the topical content of all five areas of the breadth exam is reproduced from the NCEES web-publication below:

> Depending on whether your BSCE curriculum requires all the above five technical areas, you may be wise to take some elective courses in areas not required by your program. It should be emphasized herein that the most important thing is the practical experience gained in the four years of working as a civil engineer. That is the fundamental difference between the FE and the PE exams.

All PE exams are open book. The PE Civil exam takers can also bring their own design standards/handbooks. The PE Civil passing rate for the October 2012 exam was 65% for first-time exam takers and only 27% for repeat takers (http://ncees.org/exams/pe-exam/).

## 7.6 Master's and Ph.D. Degrees

The body of knowledge in modern civil engineering is expanding significantly because of the rapid changes in globalization, information technology, diversity in society, emerging technology, public awareness of engineering impact and the needs in civil infrastructure renewal. ASCE in 2010 made a policy statement to 'support the concept of Master's degree or Equivalent as a prerequisite for licensure and the practice of civil engineering at a professional level.' (Policy Statement 465—Academic Prerequisites for Licensure and Professional Practice(2).) While the state boards in charge of professional engineer licensing has yet to embrace this concept, it is unmistakable that ASCE's policy points to the importance of a Master's degree in combination with a BSCE degree as a fundamental prerequisite to practice civil engineering. In the same policy statement, ASCE also mentioned the importance of lifelong learning, which will be discussed later in this chapter.

There are two types of Master's degrees: Master of Science and Master of Engineering. The former emphasizes theory and the latter emphasizes practice. In reality, however, most universities offer Master of Science degrees which may also emphasize practice. In general, the Master of Engineering degree is considered a terminal degree for practicing engineers. The Master of Science degree may be considered as a stepping stone to a Doctor of Philosophy (Ph.D.) degree, although many programs allow the study of a Ph.D. degree directly after a B.S. degree without a Master of Science degree. The programs for advanced degrees are indeed set up in a variety of ways in different universities.

Many universities offer a Master of Science program in general civil engineering. The program usually requires taking courses in three or more technical areas in civil engineering; some of the courses are required, some elective. Thus it expands the breadth as well as depth of the civil engineering education. Many universities offer Master of Science degree in one of the civil engineering technical areas or allowing a student to concentrate on one area. One example is the Master of Science in Environmental Engineering degree programs. While very few B.S. in Environmental Engineering degree programs are offered, many M.S. in Environmental Engineering degree programs are, especially if a department is named as civil and environmental engineering. After all a B.S. education is supposed to be broad-based, and a Master's education is to be in-depth and concentrated on a technical area in civil engineering.

Normally a Master of Engineering program in civil engineering or one of the technical areas requires an in-depth design project on a very practical subject in addition to ten to twelve courses. A Master of Science program, on the other hand, may require a thesis worthy of the equivalent of two courses in addition to ten to twelve courses. There are programs that require only course work without project or thesis. In that case if one studies full time, it is possible to graduate within two semesters or three quarters. In case of a project or thesis requirement a student is required to have an advisor or a project director to supervise the student and more often than not to assign a topic for the project or thesis.

If a Master's program of choice is offered in a university near a population center, then it is possible to study part-time while working full time as a civil engineer. The advantage of studying part-time while working is the benefit of the synergistic relationship between work and study. The work may help focus the study on courses that have immediate impact on the work. The study may open up new opportunities in present and future work. Many companies and government agencies have generous reimbursement programs that pay full or partial cost of the Master's degree. If one studies part-time, then usually it is too demanding to take more than two courses at the same time. Taking two courses a term allows one to graduate within two to three years. Some universities offer teaching or research assistantship (TA/RA) to full time Master's students with required work load up to 20 hours per week.

Pursuing a Ph.D. degree is a very different matter. The degree is not normally required by most government agencies or private firms, although both sometimes hire Ph.D. degree holders for special purposes or programs. The degree is however a prerequisite for a teaching career in higher education, which is to be described in the next section.

A Ph.D. degree program in civil engineering or specialty technical areas is designed to produce experts or specialists. In addition to course requirements, which vary greatly from one institution to another, a dissertation is required. An advisor is required too to supervise not only the dissertation research but also in general the course of study. A successful dissertation must contain new ideas, new approaches or elements of new solutions to a new or existing problem in one of the specialty technical areas in civil engineering. The advisor simply advises but may not assign a specific topic for dissertation research. The new ideas must come from the Ph.D. degree candidate. Not only a written dissertation must be produced, the candidate for the degree must successfully defend the dissertation in an oral presentation and question-and-answer meeting, which may last for hours. There usually is a committee of three to five or even more faculty members sitting on the committee. In addition to the dissertation advisor, some universities require at least one or more members from a different department to be included in the committee. They vote to decide whether the defense is successful or not.

Rarely a Ph.D. student in civil engineering supports oneself completely. The TA/RA financial assistance is usually available to Ph.D. students. Some universities even offer fellowships or scholarships to Ph.D. students that require no work. Because of the uncertainty in the time needed to complete a Ph.D. dissertation, the financial burden would be high if without any forms of financial assistant. It is also rare that a Ph.D. student work full time and study only part-time. The research on the dissertation is too demanding to spend only part-time on it; few, if any, advisors would take on a part-time Ph.D. student.

Most universities require the test score of the GRE (Graduate Records Examination) for admission to Masters or Ph.D. degree programs. It should be mentioned that some universities also offer a Doctor of Science (Sc. D.) degree program instead of a Ph.D. degree program to emphasize the practical aspect of the study.

## 7.7 A Teaching Career in Higher education

There are two different types of higher education institutes that hire Ph.D., degree holders in civil engineering or related fields: one that offers Ph.D. degrees and one offers up to Master's degrees. While both require teaching, research and services, the extent and depth of the requirement in each of the three categories are very different. Since Ph.D. programs are inseparable with research, those offering Ph.D. degree programs are inevitably research oriented and value research achievements in their faculty members. Each of the three categories of expected activities on faculty is briefly described below.

**Teaching.** The first responsibility of any higher education institute is to teach students. Teaching is conducted by full-time and part-time instructors. Part-time instructors usually are practicing engineers and teach one or two courses per term in areas they practice. Full-time instructors are often tenured or tenure-track faculty who teaches whatever amount of courses required of them in their areas of expertise. The teaching 'load' is usually one to two courses per semester in a Ph.D. degree granting institute and three to four in a Master degree granting institute. For first time teachers, it takes a lot of time to prepare for a course. A useful rule of thumb is at least three hours of preparation for every hour of classroom time. Then, there are office-hour requirements for off classroom interaction with students: usually at least one hour for every three hours of classroom time. The making of exam problems and the grading of exam papers also take time. Most exams in civil engineering programs are essay type problems that require careful grading. Teaching assistants may grade homework assignments but usually are not involved in making exam problems and grading exam papers.

The above is a sketch of the time requirement on teaching, but it does not address the increasing demand on the quality of teaching. The classical teaching method is to lecture the course content straightforwardly.

Today the lecturing maybe aided by modern audiovisual tools but more importantly teaching becomes more than just lecturing the course content. Because of the diversity in students' background and preparedness, it is increasingly important to motive and inspire students and make the lecturing interactive and easier to accept. The classical 'sink or swim' attitude toward students is no longer acceptable and the instructor is more in a partnership with the learning student than just playing judge and giving out a grade at the end of a term. All these may need the instructor to become a student again to attend teaching workshops to learn the teaching skills necessary to succeed as a teacher in a higher education institute. In Ph.D. granting universities, advising successfully Ph.D. students is valued as part of the teaching responsibilities, which takes mentorship and excellent advising and supervision.

**Research.** Research is measured by the scholarship exhibited through peer-reviewed publications. A new faculty member may be able to publish off the Ph.D. dissertation work with the advisor at the beginning but ultimately must exhibit the ability to conduct research independently from the previous advisor. While it is possible to conduct some civil engineering research by oneself without research assistants or laboratory work, most research work do need some form of assistant. That means funding or research grants are needed to support the assistants.. The ability to acquire funding from external sources is highly valued in any university, but especially in research oriented universities. The benefits of research grants also include the ability for the faculty to attend technical and professional meetings and conferences without having to rely on university funding for travel and related expenses. Abundance of research grants without adequate research output in terms of publications, however, is viewed negatively in internal peer review. It signals the weakness in not being able to complete promised research and produce quality results that are publishable.

**Services.** Most universities have faculty governance policies that require the participation of faculty in many aspects and levels of operations and decision-making. The faculty participation is in the form of committee membership at the department, college, and university levels. The demand on faculty's time varies from committee to committee. New faculty members are expected to serve at the department level and gradually move to college and university levels progressively. Some committee membership is by election and some by appointment, depending on the culture of faculty governance of the institute. Serving on committees are both an obligation and an honor.

**Getting Tenured.** The tenure system in higher education institutes in the simplest terms means that a tenured faculty cannot be fired without cause. Most universities now also have a post-tenure review system to ward off criticism that the tenure system rewards. For all practical purposes, a tenured faculty is guaranteed a lifetime employment at the institute. Thus, getting tenured is a major milestone in a teaching career.

When a new faculty is hired at the lowest faculty rank of assistant professor, he/she is given a seven-year probation period. At the end of the sixth year, the university must make a decision, unofficially called 'up or out', on either to promote the faculty to associate professor with tenure or terminate the employment at the end of the seventh year. The decision is made jointly with faculty peers and administrators according to some established procedures, which vary from one institute to another. The decision is made based on the faculty's record and peer evaluation on teaching, research and services, and sometimes external review on research and scholarship. Student evaluation of teaching effectiveness factors significantly in the decision. Even research orientated universities are reluctant to grant tenure to a faculty who could not teach effectively.

While the tenure decision is a grave one, most universities have interim reviews so that the final decision would not be abrupt and take the faculty by surprise.

**Getting Promoted.** The promotion from assistant professor to associate professor is usually granted at the time of tenure but some institute may promote someone and grant tenure later in a separate evaluation process. The promotion to full professor is another milestone in a teaching career. In a research oriented university it typically requires achieving a national or even international reputation in one's chosen field of expertise. The full professorship is normally at the top of the academic ladder, although some universities create another rank above full professor to reward the most productive and respected faculties. Another way of honoring respected faculties is to grant them chair or endowed professorships. These professorships are not another rank but are titles that carry tangible and intangible privileges.

While a teaching career seems attractive, especially considering the tenure system and flexible working hours, it by no means is an easy career. The tenure and promotion process is demanding and competitive. The flexible hours does not translate into less working hours. It is not uncommon to find a full faculty parking lot on weekends in a research oriented university. The pay is normally below that of the private companies but some nationally reputable universities have a salary structure that is very competitive. After all, to pursue a teaching career one must have a heart in teaching and enjoy the challenge in creating new knowledge through research and scholarship.

## 7.8 Lifelong Learning

The need for continuously learning new skills and knowledge in civil engineering is the same as the need to have a Master's degree as ASCE suggested. The knowledge base is changing continuously, the code and specifications governing design and construction practices are modified periodically and with shorter intervals, and the theater of application is changing from mostly domestic to international. Lifelong learning becomes necessary just to 'keep up' in the field of civil engineering. Furthermore, for a licensed professional engineer, many states now require periodical certification of continuing education credits to maintain the license. There are many forms of lifelong learning that are officially acceptable to the state boards. But in its most basic meaning, lifelong learning is the desire and the practice of self renewal in the professional sense. It may take the form of studying professional journals and reading textbooks on one's own on an advanced subject. Some of the more popular and recognized forms of lifelong learning is briefly described below.

**Attending Professional Seminars.** Most professional societies such as ASCE, ACI, AISC conduct frequent technical seminars around the country. Local chapters of professional societies also conduct technical seminars during their regular meetings and special seminars when they host prominent visitors. These seminars serve to bring the most recent developments in the profession to the local professionals.

**Attending Technical Workshops.** Professional societies sponsor technical workshops around the country when there are changes in design or construction codes and specifications. These workshops are usually very detailed with design examples using the new practice. Attending these workshops is not only helpful in one's daily practice but necessary.

**Attending Professional Society Conferences.** Professional societies held conferences once or twice annually. A variety of technical meetings are organized and participants can select the most relevant ones to attend.

**Attending Classes in Local Universities.** Sometimes it is advantageous to come back to school again and take one course to learn something new, usually a graduate level course. The rigor of a graduate level course benefits all attendants.

**Lecturing and Making Presentations.** Reporting on engineering projects in technical meetings that contain new elements of practice is another form of learning. Lecturing at undergraduate or graduate level regularly or on special occasions is also a form of lifelong learning.

## 7.9 Career Development—Management and Leadership Positions

Sooner or later opportunities present themselves for a civil engineer to move into management and leadership positions. Management is different from leadership but both require some similar personal skills and qualities. Management means a clear goal or mission is given from the leadership within an organization and the manager's clear mandate is to carry out the mission and achieve the goal. Leadership often entails the creation of a new direction or mission for a group or an organization. The required abilities for a manager and a leader are briefly described below and numbered continuous from those for a manager to those for a leader to emphasize most of the required abilities of a manger may also apply to a leader.

**I. A Manager's Abilities**

1. **Ability to Plan Ahead:** As in problem solving, a plan to achieve the given goal need to be worked out in broad outlines. This requires the understanding of the technical issues and to be up-to-date on available technical tools, and know what is possible and what is not.
2. **Ability to Organize a Team and Other Resources:** To carry out a plan, it takes people to work together as a team. The manager must know the capabilities of each team member and assign responsibilities accordingly. The manager must also know any other resources within and without the team, such as additional funding, equipment, or even personnel, and get the authority from upper management to utilize these resources. Getting additional resources is always much appreciated by team members.
3. **Ability to Articulate the Plan and Team Responsibilities Clearly:** The manager must use all the communication skills to make sure everyone on the team understands what is expected of each of them and how their responsibilities fit in the overall plan.
4. **Ability to Monitor the Progress and Change Course:** To make sure the plan is carried out timely a manager must establish milestones and monitor the progress closely but not give the impression of overbearing on the team members. Lack of progress requires the manager's attention to troubleshot and to remedy either by re-assignment of personnel or even by modifying the plan.
5. **Ability to Share Credit and Glory:** When the goal is achieved, the manager must be generous in sharing the credit with all members and should take any opportunities to promote the contribution and reputation of team members. This approach goes a long way to ensure continuous success in future goals and projects.

**II. A Leader's Abilities**

6. **Ability to Have a Clear Vision of Direction:** Whether it is for a group or the entire organization, a leader must be able to see the strengths, weaknesses, the opportunities, and the competition of the organization and establish a vision and mission for the future.
7. **Ability to Create Practical Strategies:** To achieve the vision and mission, a broad outline of the strategies must be created. This requires a complete knowledge of all the resources of the organization and

potential resources gained from being in partnership with other organizations. These resources are to be utilized in specific ways to achieve the vision and mission.

8. **Ability to Inspire and Motivate Colleagues:** A superb communication skill is a necessary condition to inspire and motivate but must be applied in combination with a clear vision and strategy to achieve results. A leader's ability to inspire and motivate tends to bring out the best from everyone.
9. **Ability to Manage the Managers:** To successfully carry out the strategies it is necessary to organize the managers and their groups to the optimum state, i.e. utilize their strengths to the fullest and avoid their weaknesses. This requires the knowledge the personal qualities of the top managers and properly put them to use.
10. **Ability to Make Timely Decisions:** Not all things turn out well all the time. Mid-course adjustment of strategies may become necessary. Only a leader can make decisions to change and the decision must be made at the most opportune time in order to have the most effect.

In addition to the above abilities, there are important qualities for a leader described below.

11. **Impeccable Personal Integrity:** A leader sets the standard of conduct of all members of an organization. The so-called 'moral authority' is derived from the daily behavior of the leader. Honesty in informing members of good and bad news is appreciated by all and the upholding of professional ethical standards to the highest degree sets the tune for the whole organization.
12. **Optimism and Positive Attitude:** if a leader does not believe in the vision and mission and the strategies designed and put forth by himself/herself, how can one expects others to believe? A leader's attitude is contagious. Positive attitude toward the outcomes of the strategies inspires others to work to achieve the outcomes.
13. **Creativity:** Find new directions and ways to achieve a mission requires 'out of the box' thinking. To see others do not see and invent new approaches is a winning quality in a leader.
14. **Fairness:** One of the central responsibilities of a leader is the periodical evaluation of performances of subordinates. Rewarding outperformers and inspiring underperformers are the necessary duties of the leader. Whatever the decision the leader makes in performance evaluation, it will be well received if it is carries out fairly and without ambiguity.
15. **Empathy:** The ability to feel and understand the emotion of others is a great asset in a leader. It helps establish a good relationship between two human beings and helps promote a healthy workplace atmosphere.
16. **Respect Others:** One of the most important attributes of a good leader is the ability to respect others. Everyone needs respect and performs the best if one feels respected.

## 7.10 Non-Civil Engineering Careers

A BSCE degree does not restrict the degree holder to a civil engineering career. As described in Chapter 1, there are several related disciplines that a civil engineering degree holder can get into. For example, a civil engineer with structural engineering interest and abilities can be hired by automobile companies, offshore platform design and construction companies, and even space industry to design structures. One can also choose to pursuit a completely different career after a BSCE degree. Some of the possibilities are described below.

1. **Medical Career.** A BSCE degree holder is qualified to apply to medical schools. Most medical schools require the test score of the Medical College Admission Test (MCAT). If admitted, the study period is normally four years. After the medical degree, there is typically a 3-5 year residency/internship depending on the specialty, and 1-3 years of fellowship for further sub-specialization if desired. Then, one needs to pass a Board exam regulated by each state to practice medicine.

   While the pay for a medical doctor is normally much higher than that of a civil engineer, considering the ten or more years required before one can begin to practice medicine and the usually demanding hours in practice, pay incentive cannot be the deciding factor in choosing a medical career. Besides, the tuition and fees for the four-year medical school study is very high and many students have to go into debt and accumulate more than $100,000 of debt in at the time of graduation. One must feel very strongly about helping people and relieving patients of the agony and suffering brought upon by illness to actually pursue a medical career.
2. **Law Career.** A BSCE degree holder is qualified to apply to law schools. The test to take is the Law School Admission Test (LSAT). If admitted, the study period is normally three years. Depending on the reputation of the law school, summer internship maybe readily available to law school students. To practice law, one must first to pass the bar examination; the difficulty to pass has been the subject of many movies and fictions. The public perception is lawyers make good money, but surprisingly the average income of lawyers is actually lower than that of civil engineers. That is because to every high profile and successful lawyer there may be ten struggling lawyers. For civil-engineer-turned-lawyers, however, there is a niche market: patent law. A patent lawyer assists inventors to apply for patent right and also assists patent holders to protect their patent right.
3. **Business Career:** After a BSCE degree one may choose to pursue a business career either by entering the business world directly or apply to enter a Master of Business Administration (MBA) degree program. The test to take is GMAT (Graduate Management Admissions Test). Some MBA programs accept previous GRE (Graduate Records Examination) test results in lieu of the GMAT score. The degree requirements vary from one program to another but it generally requires two-years of full time study or three to four years of part-time study. A combination of BSCE-MBA is very attractive to many corporations and many large civil engineering firms. Success in a business career is never guaranteed with a MBS degree, however, as there are many MBA degree holders and the business world is just as competitive if not even more competitive than the civil engineering world.
4. **Any Other Careers:** The education of a BSCE program empowers the degree holder to pursue whatever career path one chooses to follow. The many abilities and qualities needed for a civil engineering career are equally useful for any other career. In the end it is a personal choice and we encourage you to follow your heart to make career choices but use your mind to ensure success.

## Additional Reading Recommendations

1. *Policy Statement 465—Academic Prerequisites for Licensure and Professional Practice,* ASCE , 2010. http://www.asce.org/Public-Policies-and-Priorities/Public-Policy-Statements/Policy-Statement-465—Academic-Prerequisites-for-Licensure-and-Professional-Practice(2)/.
2. The Top 500 Design Firms, 2012, Engineering News-Record, http://enr.construction.com/toplists/designfirms/001-100.asp

## Assignments

1. Visit your university's career or placement center and study the resources it offers. (a) Write a summary of these resources. (b) Ask assistance from the center in writing up your own resume.
2. Visit the www.ncees.org site for the PE exam http://ncees.org/exams/pe-exam/. Download the detailed topical listing of the five civil engineering areas tested in the morning session. Associate each topical area with a course your program offers. Leave it blank if your program does not offer a course covering the topics.
3. By now you should already have some feeling on how successful you are in each of the courses you are taking. Write a one page report reviewing the success in each course and describe your plan for further effort for the reminder of the term.

# Solutions to Assignments

### Chapter 1: What is Civil Engineering?

4. (a); 5. (e); 6. (b); 7. (c); 8. (c); 9. (c); 10. (c); 11. (e); 12. (d); 13. (d); 14. (d); 15. (e)

### Chapter 2: How to Succeed?

3. The answer is 113. One arrives at this answer by following the step-by-step deduction: 3 x 5 = 15. 15 + 2 = 17. This number (17) satisfies the 3-count with 2 left. But 17 would have 2 left with 5-count (3 x 5 + 2). To arrive at 3 left with 5-count, we need to add one more but not add anything to upset the 3-count number. That means we need to add a multiple of 3 that results in adding 1 to the 5-count. The number is 6. So we arrive at 17 + 6 = 23. But it has 2 left when counting with 7 (7 x 3 + 2) while we want only 1 left. So we need to add some numbers that will not upset the 3 and 5-counts but will add 6 more for the 7-count. That number is 15 x 6 = 90 because each 15 adds 1 to the 7-count but nothing to the 3 and 5-counts. The resulting number is 23 + 90 = 113. Note you may add to 113 any multiple of 3 x 5 x 7 = 105 and it will do for the problem posted but then it will exceed the known class number of 150. You can also subtract 105 from 113 to get 8, but then it won't fit the description that many students did show up. So 113 is the only right answer.

4.(a); 5. (e); 6. (a, b, c, d); 7. (a); 8. (b); 9. (e); 10. (a, c, d); 11. (e); 12. (c); 13. (e); 14. (b); 15. (e).

### Chapter 3: The Civil Engineering Curriculum

5. (d); 8. (a); 9. (e); 12. (e); 13.(e); 14. (a); 15. (b); 16. (e); 17. (c); 18. (a).

### Chapter 4: Co-Curricular Learning

2. (b); 3. (a); 4. (a); 5. (b); 6. (d); 7. (b);8. (a);9. (b);10. (a); 11. (e); 12. (e).

### Chapter 5: Legends, Milestones and Landmarks

4. (c); 5. (c); 6. (d); 7. (d); 8. (c); 9. (a); 10. (d); 11. (c); 12. (c); 13. (a); 14. (a).

### Chapter 6: Engineering Ethics

1. (a); 2. (e); 3. (a); 4. (d); 5. (b); 6. (e); 7. (e); 8. (d); 9. (c); 10. (c).

### Chapter 7: Life beyond the BSCE Degree

NONE needed.

# Index

G



H



I



J



K



L



M

www.ingramcontent.com/pod-product-compliance
Lightning Source LLC
Chambersburg PA
CBHW081215180926
51613CB00003B/242

* 9 7 8 1 5 1 6 5 5 3 1 2 9 *